D0171785

Thinking Machines

EAGLE VALLEY LIBRARY DISTRICT
P.O. BOX 240 600 BROADWAY
EAGLE, CO 81631 / 328-8800

THINKING MACHINES

THE QUEST FOR ARTIFICIAL INTELLIGENCE—AND WHERE IT'S TAKING US NEXT

Luke Dormehl

A TarcherPerigee Book

tarcherperigee

An imprint of Penguin Random House LLC
375 Hudson Street
New York, New York 10014

Copyright © 2017 by Luke Dormehl
Penguin supports copyright. Copyright fuels creativity, encourages diverse
voices, promotes free speech, and creates a vibrant culture.
Thank you for buying an authorized edition of this book and for
complying with copyright laws by not reproducing, scanning,
or distributing any part of it in any form without permission.
You are supporting writers and allowing Penguin to continue to
publish books for every reader.

Tarcher and Perigee are registered trademarks, and the colophon is
a trademark of Penguin Random House LLC.

Most TarcherPerigee books are available at special quantity discounts for
bulk purchase for sales promotions, premiums, fund-raising, and educational
needs. Special books or book excerpts also can be created to fit specific needs.
For details, write: SpecialMarkets@penguinrandomhouse.com.

LIBRARY OF CONGRESS CATALOGING-IN-PUBLICATION DATA
Names: Dormehl, Luke, author.
Title: Thinking machines : the quest for artificial intelligence—and where it's taking us next /
Luke Dormehl.
Description: New York : TarcherPerigee, 2017. | Includes bibliographical references and index.
Identifiers: LCCN 2016044018 (print) | LCCN 2016049722 (ebook) | ISBN 9780143130581
(paperback) | ISBN 9781524704414 (ebook)
Subjects: LCSH: Artificial intelligence. | Artificial intelligence—Social aspects. | BISAC:
COMPUTERS /Intelligence (AI) & Semantics. | SCIENCE/Philosophy & Social Aspects. |
TECHNOLOGY & ENGINEERING / History.
Classification: LCC TA347.A78 D67 2017 (print) | LCC TA347.A78 (ebook) | DDC
006.309—dc23
LC record available at https://lccn.loc.gov/2016044018

Printed in the United States of America
1   3   5   7   9   10   8   6   4   2

Book design by Katy Riegel

While the author has made every effort to provide accurate telephone numbers, Internet
addresses, and other contact information at the time of publication, neither the publisher nor
the author assumes any responsibility for errors or for changes that occur after publication.
Further, the publisher does not have any control over and does not assume any responsibility
for author or third-party Web sites or their content.

To my pal

Alex Millington

CONTENTS

INTRODUCTION

Thinking Machines

THE ALL-SEEING RED eye stares, unblinking. The computer speaks calmly.

"Hello," it says. "Shall we continue the game?"

It is referring to a game of chess you started with it earlier that day. But you're not really in the mood to play. It's not that the computer almost always beats you at chess (although it does). Instead, you're annoyed because it made an inexplicable error concerning the supposed failure of vital bits of important machinery, necessary to ensure your continued survival. *No biggie.* You checked them out in person and found them to still be in good working order, although the computer insisted they were broken. Now you want answers.

"Yes, I know that you found them to be functional, but I can assure you that they were *about* to fail," the machine says, trying to placate you in the same emotionless monotone it always uses.

You can feel your blood boiling.

"Well, that's just not the case," you splutter. "The components are perfectly all right. We tested them under 100 percent overload conditions."

"I'm not questioning your word, but it's just not possible," the computer says.

It then adds the six words you know to be true, but are absolutely the last thing you want to hear right now: "I'm not capable of being wrong."

Movie buffs will instantly recognize this scene from Stanley Kubrick's classic sci-fi film, *2001: A Space Odyssey*, about a sentient computer that turns murderous and begins attempting to kill off its crew.

For years, this was the way we thought about Artificial Intelligence: as a faintly threatening presence safely shrouded in the context of science fiction.

No more.

Today, the dream of AI has stepped out of cinemas and paperback novels and into reality. Artificial Intelligence can drive cars, trade stocks and shares, learn to carry out complex skills simply by watching YouTube videos, translate across dozens of different languages, recognize human faces with more accuracy than we can, and create original hypotheses to help discover new drugs for curing disease. That's just the beginning.

Thinking Machines is a book about this journey—and what it means for all of us. En route, we will meet computers pretending to trap pedophiles, dancing robot vacuum cleaners, chess-playing algorithms, and uploaded consciousnesses designed to speak to you from beyond the grave. This is the story of how we imagine our future, and how in a world obsessed with technology, we carve out a role for humanity in the face of accelerating computer intelligence. It's about the nature of creativity, the future of employment, and what happens when all knowledge is data and can be stored electronically. It's about what we're trying to do when we make machines smarter than we are, how humans still have the edge (for now), and the question of whether you and I aren't thinking machines of a sort as well.

The pioneering British mathematician and computer scientist Alan Turing predicted in 1950 that by the end of the twentieth century, "the use of words and general educated opinion will have altered so much that one will be able to speak of machines thinking without expecting to be contradicted."

Like many futurist predictions about technology, he was optimistic in his timeline—although he wasn't off by too much. In the early part of the twenty-first century, we routinely talk about "smart" connected technologies and machine "learning"— concepts that would have seemed bizarre to many people in Turing's day.

Now celebrating its sixtieth year as a discipline, Artificial Intelligence is cementing itself as one of mankind's biggest and most ambitious projects: a struggle to build real thinking machines. Technologists are getting closer by the day, and a glimmering tomorrow is fast coming into focus on the horizon.

Thinking Machines is about this dazzling (near) future, the changes that lurk just around the corner, and how they will transform our lives forever.

1

Whatever Happened to Good Old-Fashioned AI?

IT WAS THE first thing people saw as they drew close: a shining, stainless steel globe called the Unisphere, rising a full twelve stories into the air. Around it stood dozens of fountains, jetting streams of crystal-clear water into the skies of Flushing Meadows Corona Park, in New York's Queens borough. At various times during the day, a performer wearing a rocket outfit developed by the US military jetted past the giant globe—showing off man's ability to rise above any and all challenges.

The year was 1964 and the site, the New York World's Fair. During the course of the World's Fair, an estimated 52 million people descended upon Flushing Meadows' 650 acres of pavilions and public spaces. It was a celebration of a bright present for the United States and a tantalizing glimpse of an even brighter future: one covered with multilane motorways, glittering skyscrapers, moving pavements and underwater communities. Even the possibility of holiday resorts in space didn't seem out of reach for a country like the United States, which just five years later would successfully send man to the Moon. New York City's "Master Builder" Robert Moses referred to the 1964 World's Fair as "the Olympics of Progress."

Wherever you looked there was some reminder of America's post-war global dominance. The Ford Motor Company chose the World's Fair to unveil its latest automobile, the Ford Mustang, which rapidly became one of history's best-selling cars. New York's Sinclair Oil Corporation exhibited "Dinoland," an animatronic recreation of the Mesozoic age, in which Sinclair Oil's brontosaurus corporate mascot towered over every other prehistoric beast. At the NASA pavilion, fairgoers had the chance to glimpse a fifty-one-foot replica of the Saturn V rocket ship boat-tail, soon to help the Apollo space missions reach the stars. At the Port Authority Building, people lined up to see architects' models of the spectacular "Twin Towers" of the World Trade Center, which was set to break ground two years later in 1966.

Today, many of these advances evoke a nostalgic sense of technological progress. In all their "bigger, taller, heavier" grandeur, they speak to the final days of an age that was, unbeknownst to attendees of the fair, coming to a close. The Age of Industry was on its way out, to be superseded by the personal computer–driven Age of Information. For those children born in 1964 and after, digits would replace rivets in their engineering dreams. Apple's Steve Jobs was only nine years old at the time of the New York World's Fair. Google's cofounders, Larry Page and Sergey Brin, would not be born for close to another decade; Facebook's Mark Zuckerberg for another ten years after that.

As it turned out, the most forward-looking section of Flushing Meadows Corona Park turned out to be the exhibit belonging to

International Business Machines Corporation, better known as IBM. IBM's mission for the 1964 World's Fair was to cement computers (and more specifically Artificial Intelligence) in the public consciousness, alongside better-known wonders like space rockets and nuclear reactors. To this end, the company selected the fair as the venue to introduce its new System/360 series of computer mainframes: machines supposedly powerful enough to build the first prototype for a sentient computer.

IBM's centerpiece at the World's Fair was a giant, egg-shaped pavilion, designed by the celebrated husband and wife team of Charles and Ray Eames. The size of a blimp, the egg was erected on a forest of forty-five stylized, thirty-two-foot-tall sheet metal trees; a total of 14,000 gray and green Plexiglas leaves fanning out to create a sizable, one-acre canopy. Reachable only via a specially installed hydraulic lift, the egg welcomed in excited fair attendees so that they could sit in a high-tech screening room and watch a video on the future of Artificial Intelligence. "See it, THINK, and marvel at the mind of man and his machine," wrote one giddy reviewer, borrowing the "Think" tagline that had been IBM's since the 1920s.

IBM showed off several impressive technologies at the event. One was a groundbreaking handwriting recognition computer, which the official fair brochure referred to as an "Optical Scanning and Information Retrieval" system. This demo allowed visitors to write an historical date of their choosing (post-1851) in their own handwriting on a small card. That card was then fed into an "optical character reader," where it was converted into digital form, and then relayed once more to a state-of-the-art IBM 1460 computer system. Major news events were stored on disk in a vast database and the results were then printed onto a commemorative punch-card for the amazement of the user. A surviving punch-card reads as follows:

THE FOLLOWING NEWS EVENT WAS REPORTED IN
THE NEW YORK TIMES ON THE DATE THAT YOU
REQUESTED:

APRIL 14, 1963: 30,000 PILGRIMS VISIT JERUSALEM
FOR EASTER; POPE JOHN XXIII PRAYS FOR TRUTH &
LOVE IN MAN.

Should a person try and predict the future—as, of course, some wag did on the very first day—the punch-card noted: "Since this date is still in the future, we will not have access to the events of this day for [insert number] days."

Another demo featured a mechanized puppet show, apparently "fashioned after eighteenth-century prototypes," depicting Sherlock Holmes solving a case using computer logic.

Perhaps most impressive of all, however, was a computer that bridged the seemingly unassailable gap between the United States and Soviet Union by translating effortlessly (or what appeared to be effortlessly) between English and Russian. This miraculous technology was achieved thanks to a dedicated data connection between the World's Fair's IBM exhibit and a powerful IBM mainframe computer 114 miles away in Kingston, New York, carrying out the heavy lifting.

Machine translation was a simple, but brilliant, summation of how computers' clear-thinking vision would usher us toward utopia. The politicians may not have been able to end the Cold War, but they were only human—and with that came all the failings one might expect. Senators, generals and even presidents were severely lacking in what academics were just starting to call "machine intelligence." Couldn't smart machines do better? At the 1964

World's Fair, an excitable public was being brought up to date on the most optimistic vision of researchers. Artificial Intelligence brought with it the suggestion that, if only the innermost mysteries of the human brain could be eked out and replicated inside a machine, global harmony was somehow assured.

Nothing summed this up better than the official strapline of the fair: "Peace Through Understanding."

Predicting the Future

Two things stand out about the vision of Artificial Intelligence as expressed at the 1964 New York World's Fair. The first is how bullish everyone was about the future that awaited them. Despite the looming threat of the Cold War, the 1960s was an astonishingly optimistic decade in many regards. This was, after all, the ten-year stretch that began with President John F. Kennedy announcing that, within a decade, man would land on the moon—and ended with exactly that happening. If that was possible, there seemed no reason why unraveling and re-creating the mind should be any tougher to achieve. "Duplicating the problem-solving and information-handling capabilities of the [human] brain is not far off," claimed political scientist and one of AI's founding fathers, Herbert Simon, in 1960. Perhaps borrowing a bit of Kennedy-style gauntlet-throwing, he casually added his own timeline: "It would be surprising if it were not accomplished within the next decade."

Simon's prediction was hopelessly off, but as it turns out, the second thing that registers about the World's Fair is that IBM wasn't wrong. All three of the technologies that dropped jaws in 1964 are commonplace today—despite our continued insistence that AI is not yet here. The Optical Scanning and Information Retrieval has become the Internet: granting us access to more information at a moment's notice than we could possibly hope to absorb in a lifetime. While we still cannot see the future, we are making enormous advances in this capacity, thanks to the huge data sets generated by users that offer constant forecasts about the news stories, books or songs that are likely to be of interest to us. This predictive connectivity isn't limited to what would traditionally be thought of as a computer, either, but is embedded in the devices, vehicles and buildings around us thanks to a plethora of smart sensors and devices.

The Sherlock Holmes puppet show was intended to demonstrate how a variety of tasks could be achieved through computer logic. Our approach to computer logic has changed in some ways, but Holmes may well have been impressed by the modern facial recognition algorithms that are more accurate than humans when it comes to looking at two photos and saying whether they depict the same person. Holmes's creator, Arthur Conan Doyle, a trained doctor who graduated from Edinburgh (today the location of one of the UK's top AI schools), would likely have been just as dazzled by Modernizing Medicine, an AI designed to diagnose diseases more effectively than many human physicians.

Finally, the miraculous World's Fair Machine Translator is most familiar to us today as Google Translate: a free service that offers impressively accurate probabilistic machine translation between some fifty-eight different languages—or 3,306 separate translation services in total. If the World's Fair imagined instantaneous translation between Russian and English, Google Translate goes further still by also allowing translation between languages like Icelandic and Vietnamese, or Farsi and Yiddish, which have had historically limited previous translations. Thanks to cloud computing, we don't even require stationary mainframes to carry it out, but rather portable computers, called smartphones, no bigger than a deck of cards.

In some ways, the fact that all these technologies now exist—not just in research labs, but readily available to virtually anyone who wants to use them—makes it hard to argue that we do not yet live in a world with Artificial Intelligence. Like many of the shifting goalposts we set for ourselves in life, it underlines the way that AI represents computer science's Neverland: the fantastical "what if" that is always lurking around the next corner.

With that said, anyone thinking that the development of AI from its birth sixty years ago to where it is today is a straight line is very much mistaken. Before we get to the rise of the massive "deep learning neural networks" that are driving many of our most notable advances in the present, it's important to understand a bit more about the history of Artificial Intelligence.

And how, for a long time, it all seemed to go so right before going wrong.

The Giant Brain

The dream of bringing life to inanimate objects has been with us for thousands of years. However, when it comes to the popularization of Artificial Intelligence for regular people, it makes sense to begin with the world's first programmable computer: a thirty-ton colossus named ENIAC. Powered on at the University of Pennsylvania just six months after the Second World War ended in 1945, ENIAC stood for Electronic Numeric Integrator and Calculator. It had cost $500,000 of US military funding to create and possessed a speed that was around 1,000 times faster than other electromechanical machines it may have competed against. The machine, and the idea that it represented, fascinated the press. They took to calling it "the giant brain."

The notion of building such a "giant brain" captured the popular imagination. Until the end of the Second World War, a "computer" was the term used for a person who carried out calculations in a field such as bookkeeping. All of a sudden, computers were no longer people, but machines equipped with vacuum tubes and transistors—yet capable of performing calculations at a speed even greater than the most gifted of people. The Second World War and its immediate aftermath triggered a surge of interest in the field of cognitive psychology. During wartime alone, membership of the American Psychological Association expanded from 2,600 to 4,000. By 1960—fifteen years later—it would hit 12,000 members. Researchers in cognitive psychology imagined the human brain itself as a machine, from which complex behavior arose as the aggregate result of multiple simple responses. Instead of wasting their time on unprovable "mental entities," cognitive psychologists focused their attention only on what was strictly observable about human behavior. This was the birth of fields like "behaviorism," which the influential psychologist B. F. Skinner (known for his experiments with rats) described as the "technology of behavior."

Engineers may previously have balked at the more metaphysical aspects of psychology, but they were intrigued at the concept that the brain might be a computer. They were equally fascinated by the new focus on understanding memory, learning and reasoning, which many psychologists felt were the basis for human intelligence. Excitingly, they also saw the potential advantages machines had over people. ENIAC, for instance, could perform an astonishing 20,000 multiplications per minute. Compared with the unreliable memory of humans, a machine capable of accessing thousands of items in the span of microseconds had a clear advantage.

There are entire books written about the birth of modern computing, but three men stand out as laying the philosophical and technical groundwork for the field that became known as Artificial Intelligence: John von Neumann, Alan Turing and Claude Shannon.

A native of Hungary, von Neumann was born in 1903 into a Jewish banking family in Budapest. In 1930, he arrived at Princeton University as a math teacher and, by 1933, had established himself as one of six professors in the new Institute for Advanced Study in Princeton: a position he stayed in until the day he died. By any measure, von Neumann was an astonishing intellect. According to legend, he was able to divide eight-digit numbers in his head at the age of six. During the Second World War, von Neumann worked on the Manhattan Project at Los Alamos, where one of his jobs was the terrible task of working out the precise height at which the hydrogen bomb must explode to cause maximum devastation. Von Neumann's major contribution to computing was helping to establish the idea of a computer program store in the computer memory. Von Neumann was, in fact, the first person to use the human terminology "memory" when referring to a computer. Unlike some of his contemporaries, he did not believe a computer would be able to think in the way that a human can, but he did help establish the parallels that exist with human physiognomy. The parts of a computer, he wrote in one paper, "correspond to the associative neurons in the human nervous system. It remains to discuss the equivalents of the sensory or afferent and the motor or efferent neurons." Others would happily take up the challenge.

Alan Turing, meanwhile, was a British mathematician and cryptanalyst. During the Second World War, he led a team for the Government Code and Cypher School at Britain's secret code-breaking center, Bletchley Park. There he came up with various techniques for cracking German codes, most famously an electro-mechanical device capable of working out the settings for the Enigma machine. In doing so, he played a key role in decoding in-tercepted messages, which helped the Allies defeat the Nazis. Tur-ing was fascinated by the idea of thinking machines and went on to devise the important Turing Test, which we will discuss in detail in a later chapter. As a child, he read and loved a book called *Natural Wonders Every Child Should Know,* by Edwin Tenney Brewster, which the author described as "an attempt to lead children of eight or ten, first to ask and then to answer the question: 'What have I in common with other living things, and how do I differ from them?'" In one notable section of the book, Brewster writes:

> Of course, the body is a machine. It is a vastly complex ma-chine, many, many times more complicated than any other machine ever made with hands; but after all a machine. It has been likened to a steam engine. But that was before we knew as much about the way it works as we know now. It really is a gas engine: like the engine of an automobile, a motor boat, or a flying machine.

One of Turing's most significant concepts related to something called the Universal Turing Machine. Instead of computers being single-purpose machines used for just one function, he explained how they could be made to perform a variety of tasks by reading step-by-step instructions from a tape. By doing so, Turing wrote that the computer "could in fact be made to work as a model of any other machine." This meant that it was not necessary to have infinite different machines carrying out different tasks. As Turing noted, "The engineering problem of producing various machines for various jobs is replaced by the office work of 'programming' the universal machine to do these jobs."

One such job, he hypothesized, was mimicking human intelligence. In one notable paper, entitled "Intelligent Machinery," Turing considered what it would take to reproduce intelligence inside a machine: a particular challenge given the limitations of computers at the time. "The memory capacity of the human brain is probably of the order of ten thousand million binary digits," he considered. "But most of this is probably used in remembering visual impressions, and other comparatively wasteful ways. One might reasonably hope to be able to make some real progress [toward Artificial Intelligence] with a few million digits [of computer memory]."

The third of AI's forefathers was a man named Claude Shannon, known today as the father of "information theory." Born in 1916—making him the youngest of the three—Shannon's big contribution to computing related to the way in which transistors work. Transistors are the billions of tiny switches that make up a computer. An algorithm is the sequence of instructions that tells a computer what to do by switching these transistors on and off. By having certain transistors switch on and off in response to other transistors, Shannon argued that computers were performing basic reasoning. If, he said, transistor 1 switches on when transistors 2 and 3 are also on, this is a logical operation. Should transistor 1 turn on when either transistor 2 or 3 is on, this is a second logical operation. And if transistor 1 turns on when transistor 2 is switched off, this is a third logical operation. Like a simple vocabulary of spoken language, all computer algorithms break down into one of three different states: AND, OR, and NOT. Combining these simple states into complex series of instructions, Shannon suggested that complex chains of logical reasoning could be carried out.

The Dartmouth Conference

Of this group, only Shannon went on to play an active role in the official formation of Artificial Intelligence as its own discipline. Both Turing and von Neumann died tragically young, aged just forty-one and fifty-three respectively, although their ideas and influence continue to be felt today. Alan Turing was a homosexual at a time in English history in which it was a crime to be so. Despite his code-breaking work being vital to the British war effort against Nazi Germany, he was prosecuted and convicted of gross indecency in 1952. Forced to choose between prison and a painful chemical castration process, Turing opted for the latter. Two years later, he committed suicide by taking a bite of an apple laced with cyanide. He was given a posthumous royal pardon in 2013, and the suggestion was made that a "Turing's Law" should be passed to pardon other gay men historically convicted of indecency charges.

Von Neumann's death was caused by cancer, quite possibly the result of attending nuclear tests as part of the atom bomb project. In his obituary in the *Economic Journal*, one of von Neumann's close colleagues described his mind as "so unique that some people have asked themselves—they too eminent scientists—whether he did not represent a new stage in human mental development."

With two of its founders gone, the growing interest in building thinking machines was picked up by other, younger researchers. AI's second wave of researchers became the first to officially name the field: formalizing it as its own specialized discipline. In the summer of 1956—when Elvis Presley was scandalizing audiences with his hip gyrations, Marilyn Monroe married playwright Arthur Miller, and President Dwight Eisenhower authorized "In God we trust" as the US national motto—AI's first official conference took place. A rolling six-week workshop, bringing together the smartest academics from a broad range of disciplines, the event unfolded on the sprawling 269-acre estate of Dartmouth College in Hanover, New England. Along with Claude Shannon, two of the organizers were young men named John McCarthy and Marvin Minsky, both of whom became significant players in the growing field of Artificial Intelligence.

"The study [of AI] is to proceed on the basis of the conjecture that every aspect of learning or any other feature of intelligence can be so precisely described that a machine can be made to simulate it," they wrote. "An attempt will be made to find how to make machines use language, form abstractions and concepts, solve the kinds of problems now reserved for humans, and improve themselves."

Their ambition and self-belief was absolute, but their timeframe was perhaps somewhat compressed. "We think a significant advance can be made in one or more of these problems if a carefully selected group of scientists work on it for a summer," they argued in their proposal for the Dartmouth conference.

Needless to say, things took a bit longer than that.

Look, Ma, No Hands!

As more researchers took an interest in AI, it began to subdivide into different fields, reflecting the massive scope of what was being attempted. In some senses, this was inevitable. At the Dartmouth conference, it had proven difficult to even get everyone to agree on a name for their new field. John McCarthy pushed for the flashy-sounding Artificial Intelligence. Others were less convinced. Another researcher named Arthur Samuel thought the name sounded "phony," while still others—Alan Newell and Herbert Simon—immediately reverted to calling their work "complex information programming."

The rapid division of Artificial Intelligence into different specialties didn't take long. For evidence, look no further than the UK's "Mechanization of Thought Processes" conference, organized at the National Physical Laboratory in Teddington, Middlesex, in 1958. Just two years after the Dartmouth conference, AI was already split into fields including "artificial thinking, character and pattern recognition, learning, mechanical language translation, biology, automatic programming, industrial planning and clerical mechanization."

The period that followed is often considered to be the glory days of classic AI. The field was fresh, apparent progress was being made, and thinking machines seemed to lurk just over the horizon. It didn't hurt that funding was plentiful, either—largely thanks to government organizations such as the US Defense Department's Advanced Research Projects Agency (ARPA). In June 1963, ARPA issued MIT a $2.2 million grant for researching "machine-aided cognition." According to people who benefited from the grant,

ARPA paid it in one installment and didn't show much concern for how it was spent. This was far from an isolated incident.

John McCarthy referred to this as the "Look, Ma, no hands!" era of AI, summoning up images of youthful exuberance and derring-do. For more than a decade, researchers attacked an astonishingly broad series of challenges, often designed to disprove a hypothesis like: "No machine will ever be capable of doing X." What the X stood for depended on who was doing the asking. One researcher wrote a checkers program capable of beating most amateurs, including himself. Another breakthrough included a perceptive AI able to rearrange colored, differently shaped blocks on a table using a robotic hand: an astonishing feat in computer vision. A program called SAINT proved able to solve calculus integration problems of the level found on a first-year college course. Another, called ANALOGY, did the same for the geometric questions found in IQ tests, while STUDENT cracked complex algebra story conundrums such as: "If the number of customers Tom gets is twice the square of 20 percent of the number of advertisements he runs, and the number of advertisements he runs is 45, what is the number of customers Tom gets?"*

A particularly impressive display of computational reasoning was a program called SIR (standing for Semantic Information Retrieval). SIR appeared to understand English sentences and was even able to learn relationships between objects in a way that resembled real intelligence. In reality, this "knowledge" relied on a series of pre-programmed templates, such as A is a part of B, with nouns substituting for the variables. However, it was enough to suggest to the likes of Marvin Minsky that similar approaches

---

*The answer, in case you want to prove yourself as smart as an AI, is 162.

could begin to tackle a variety of problems. Here is a typical example of SIR in action, with the lowercase sentences referring to the
text entered by the user and the uppercase sentences referring to
answers by the machine:

> Every boy is a person.
> I UNDERSTAND.

> A finger is part of a hand.
> I UNDERSTAND.

> Each person has two hands.
> THE ABOVE SENTENCE IS AMBIGUOUS. PLEASE
> REPHRASE IT.

> There are two hands on each person.
> I UNDERSTAND.

> How many fingers does John have?
> THE ABOVE SENTENCE IS AMBIGUOUS, BUT I
> ASSUME (HAS) MEANS (HAS AS PARTS). I DON'T
> KNOW WHETHER FINGER IS PART OF JOHN.

> John is a boy.
> I UNDERSTAND.

> How many fingers does John have?
> THE ABOVE SENTENCE IS AMBIGUOUS, BUT I
> ASSUME (HAS) MEANS (HAS AS PARTS). HOW MANY
> FINGERS PER HAND?

Every hand has five fingers.
I UNDERSTAND.

How many fingers does John have?
THE ANSWER IS TEN.

What bonded together all of these seemingly disparate projects, all off in their own corners of AI, was the way that they conceived of intelligence and intelligent behavior. They took a top-down
view of intelligence, which has come to be known as Symbolic AI
or, today, Good Old-Fashioned AI. To a Good Old-Fashioned AI
researcher, all intelligence is based on humans' ability to understand the world by forming internal symbolic representations. We
then create rules for dealing with these concepts, and these rules
can be formalized in a way that captures everyday knowledge. If
the brain is indeed a computer, this means that every situation we
navigate relies on us running an internal computer program telling us, step by step, how to carry out an operation based entirely
on logic. And if that is the case, surely those same rules about the
organization of the world could also be passed on to a computer.

It all sounded almost too easy and, for a while, it was exactly
that.

SHAKEY in Space

Although few saw it coming, there were several problems with
Artificial Intelligence as it was developing. As is often the case
with an exciting field that resonates with the general public, part
of the blame must lie with the press. Overenthusiasm meant that

impressive, if incremental, advances were often written up as though truly smart machines were already here. For example, one heavily hyped project was a 1960s robot called SHAKEY, described as the world's first general-purpose robot capable of reasoning about its own actions. In doing so, it set benchmarks in fields like pattern recognition, information representation, problem solving and natural language processing.

That alone should have been enough to make SHAKEY exciting, but journalists couldn't resist a bit of embellishment. As such, when SHAKEY appeared in *Life* magazine in 1970, he was hailed not as a promising combination of several important research topics, but as the world's "first electronic person." Tying SHAKEY into the space mania still carrying over from the previous year's moon landing, *Life*'s reporter went so far as to claim SHAKEY could "travel about the Moon for months at a time without a single beep of direction from the earth."

This was completely untrue, although not all researchers could resist playing up to it. At an AI conference in Boston during the 1970s, one researcher told a member of the press that it would take just five more years until intelligent robots like SHAKEY were picking up the stray socks in people's homes. Pulled aside by a furious younger colleague, the researcher was told, "Don't make those predictions! People have done this before and gotten into trouble. You're underestimating how long this will take." Without pausing, the older researcher responded, "I don't care. Notice all the dates I've chosen were after my retirement date."

AI practitioners weren't always this cynical, but many were prone to the same fits of cyberbole. In 1965, Herbert Simon stated that in just twenty years' time, machines would be capable "of doing any work a man can do." Not long after, Marvin Minsky added that "within a generation . . . the problem of creating Artificial Intelligence will substantially be solved."

The Chinese Room

Philosophical problems were also beginning to be raised concerning Symbolic AI. Perhaps the best-known criticism is the thought experiment known as "the Chinese Room." Put forward by the American philosopher John Searle, it questions whether a machine processing symbols can ever truly be considered intelligent.

Imagine, Searle says, that he is locked in a room and given a collection of Chinese writings. He is unable to speak or write Chinese, and can't even distinguish Chinese writing from Japanese writing or meaningless squiggles. In the room, Searle discovers a set of rules showing him a set of symbols that correspond with other symbols. He is then given "questions" to "answer," which he does by matching the question symbols with the answer ones. After a while, Searle becomes good at this task—although he still has no concept of what the symbols are that he is manipulating. Searle asks whether it can be said that the person in the room "understands" Chinese. His answer is no, because there is a total lack of intentionality on his part. He writes: "Such intentionality as computers appear to have is solely in the minds of those who program them and those who use them, those who send in the input and those who interpret the output."

If Searle was accusing AI researchers of acting like parents willing to seize on anything to proclaim their children's brilliance, then AI researchers were, themselves, facing a similar uncomfortable truth: that their kids weren't actually all that smart. Worryingly, tools which had shown promise in lab settings proved altogether less adept at coping in real-world situations. Symbolic AI was about building top-down, rule-based systems, able to work perfectly in laboratory settings where every element could be controlled. These "micro-worlds" contained very few objects and, as a result, limited actions that could be taken. Transferred to the chaos of everyday life, programs that had worked perfectly in training froze up like the England team in a World Cup opener.

Researchers acknowledged these weaknesses, describing such micro-worlds as "a fairyland in which things are so simplified that almost every statement about them would be literally false if asserted about the real world." In all, AI struggled to deal with ambiguity; it was lacking the flexible abstract reasoning, data and processing power it needed to make sense of what it was shown. Anything that hadn't been explicitly accounted for beforehand was cause for abject panic. The American writer Joseph Campbell quipped that this form of AI was not dissimilar to Old Testament gods, with "lots of rules and no mercy."

Moravec's Paradox

Capping all of this uncertainty off was a bigger question about whether AI researchers were going about their work in the right way. A bit like starting work on a puzzle by piecing together the most complex pieces first, AI researchers had imagined that if they could solve the more advanced problems, the easy ones would take care of themselves. After all, if you can get a machine to play chess like a math prodigy, how tough could it be to simulate the learning of an infant? Pretty tough, it transpired. As a game, chess consists of clearly defined states, board positions and legal or illegal moves. It is a static world in which players have access to complete information, just so long as they can see the board and know the moves available to them. Chess may be a part of reality, but reality itself is nothing like chess. Suddenly, researchers like Hans Moravec began to voice startling suggestions like the notion that it is "comparatively easy to make computers exhibit adult-level performance on intelligence tests or playing checkers, and difficult or impossible to give them the skills of a one-year-old when it comes to perception and mobility."

This concentration on the more complex aspects of life to the exclusion of more commonplace tasks may have had something to do with the sorts of people working in AI. In many cases brilliant scientists for whom the word "prodigy" can readily be applied, these researchers could handle the minutiae of chess or Boolean logic, but were absentminded and lacking in real-life common sense. In one commonly told anecdote, a highly intelligent MIT researcher named Seymour Papert once left his wife behind at a New York airport. He only realized that she was not accompanying him when he was halfway across the Atlantic. John McCarthy, meanwhile, could be tenacious when a problem challenged him, but caused no shortage of headaches by continually forgetting to fill out progress reports for the various agencies that funded him. McCarthy's Introduction to Artificial Intelligence course at Stanford was reportedly so unfocused that students took to calling it "Uncle John's Mystery Hour" behind his back. In the way that dogs are said to resemble their owners, is it any surprise that the focus of these researchers' AI programs tended to be on lofty goals rather than mundane (but potentially more useful) feats?

As the psychologist Steven Pinker summed it up: "The main

lesson of [the first] thirty-five years of AI research is that the hard problems are easy and the easy problems are hard."

Changing Ambitions

Facing these kinds of challenges, Good Old-Fashioned AI started to run into problems. From the 1970s, the field cooled off as the optimism of previous decades dissipated. Budgets were brutally slashed, plunging Artificial Intelligence into the first of several so-called "AI Winters." In the United States, even the lovable SHAKEY the robot project shuddered to a halt when it became clear that it was not the robotic James Bond spy its funders at the Defense Department had hoped for. Forget spying, SHAKEY couldn't even replace regular troops on the battlefield! One researcher who worked on the project remembers some military types coming in for a last-ditch look at SHAKEY rolling around the laboratory at its research institute, SRI International. Turning to one of its creators, a skeptical general asked, "Would it be possible to mount a thirty-six-inch bayonet on it?"

AI responded by shifting its ambitions, scaling back on some of its grander missions in favor of narrow, well-defined problems for which clear measures of success could be made. One such area was the growing field of video games. AI had been associated with game-playing since its earliest days, when Alan Turing and Claude Shannon attempted to build an automated chess player. In that instance, chess had been a micro-world designed to prove intelligent behavior that could later be rolled out in the real world. Now video games presented an end goal in and of themselves.

Not only were researchers' skills in demand, but there was real money on offer, too. One such beneficiary was Alexey Pajitnov, a twenty-eight-year-old AI researcher then working for the Soviet Academy of Sciences' Computer Center in Moscow. In June 1984, Pajitnov created a simple program to test out the lab's new computer system. Brought to market by a shrewd entrepreneur under the name *Tetris*, Pajitnov's falling blocks game proceeded to sell more than *170 million* copies worldwide.

As the 1980s wore on, video games became increasingly intricate and AI experts were snapped up to help. Their ability to model complex behavior using simple rules meant that computer-controlled characters could possess their own motivations. In the hit game *Theme Park*, for instance, AI simple agents flocked around the parks built by users, taking routes no programmer explicitly mapped out.

In one sense, video games were the perfect place for Good Old-Fashioned AI. Questions about whether behavior was truly intelligent, or just acting like it, meant nothing if the AI was being used to model the zombie enemy in a first-person shooter. (In fact, it would be considerably crueler if the agents *were* intelligent.) Even today, video-game developers employ more AI practitioners than any other industry.

Expert Systems

A second new application for AI was working alongside humans as problem-solving tools. Although reasoning is a key part of intelligence, researchers knew that this was not the only part. To build Artificial Intelligence capable of being used in the real world to solve genuine problems, experts decided they needed machines that could combine reasoning with knowledge. For example, a computer that was going to be useful in neuroscience would have to be intimately acquainted with the same concepts, facts, representations, methods, models, metaphors and other facets of the subject that a qualified neuroscientist would be.

This meant that programmers suddenly had to become "knowledge engineers," capable of taking human experts in a variety of fields and distilling their knowledge into rules a computer could follow. The resulting programs were called "expert systems." These were systems built on an extensive collection of probabilistic "IF . . . THEN" rules. One early attempt at an expert system was called DENDRAL, a program designed to help organic chemists identify unknown organic molecules. "For a while, we were regarded at arm's length by the rest of the AI world," creator Edward Feigenbaum told Pamela McCorduck, one of the earliest writers to chronicle the history of Artificial Intelligence. "I think they thought DENDRAL was something a little dirty to touch because it had to do with chemistry, though people were pretty generous about 'oohs' and 'ahs' because it was performing like a PhD in chemistry."

Another similar project was MYCIN, designed to help recommend the correct dosage of antibiotics for severe infections such as meningitis. Like a real doctor, MYCIN drew conclusions by combining pieces of probabilistic evidence from the previous experience of its programmers. These years of experience were squeezed and shaped until they resembled "rules" like the following:

IF . . . the infection which requires therapy is meningitis, and the type of infection is fungal, and organisms were not seen on the stain of the culture, and the patient is not a com-

promised host, and the patient has been to an area that is
endemic for coccidiomycoses, and the race of the patient is
Black, Asian, or Indian, and the cryptococcal antigen in the
csf test was not positive, THEN ... there is a 50 percent
chance that cryptococcus is *not* one of the organisms which
is causing the infection.

On their own, such probabilistic rules didn't amount to much.
When combined in their hundreds, however, they could regularly
find the right answer. DENDRAL and MYCIN remained lab experiments that were never used in the real world. Another expert
system called XCON proved more successful. Created in 1978,
XCON lacked the world-improving ambitions of DENDRAL and
MYCIN. Instead of helping scientists form hypotheses, or doctors
treat infectious diseases, XCON aided engineers in configuring
VAX supercomputers by choosing the right system components
for a customer's requirements. In short, it was the world's greatest
know-it-all shop assistant.

For the first time, big business began to show a real interest in
AI as something more than a demo of the future. As long as expert
systems could make them money, it shockingly turned out that
companies didn't care too much about whether expert systems
were *real* AI or simply "clever programming." XCON's first day of
work took place in 1980 at the Salem, New Hampshire, factory of
DEC, the Digital Equipment Corporation. By 1986, XCON had
processed a whopping 80,000 orders, was saving DEC an estimated $25 million a year, and achieved accuracy rates of 95–98
percent. If it had only married the boss's daughter, it could've had
a future as CEO.

Other rival companies soon crawled out of the woodwork, offering custom solutions for companies wanting their own expert systems. Dipmeter Advisor could advise with the analysis of geological formations in oil-well drilling. The scintillating Grain Marketing Advisor made clear its ambitions to help farmers properly market and store their grain crops. "How can you take immediate advantage of expert systems technology to enhance your existing data processing applications, on your existing hardware, using your current . . . staff?" asked an ad printed in *Computerworld* magazine in October 1986. "Only Teknowledge has the answer. And it's yours. Free. At a half-day seminar in your area."

In all, during 1985, a massive $1 billion was spent by approximately 150 companies wanting to get in on the Artificial Intelligence business. That year, a meeting of the American Association for Artificial Intelligence and the International Joint Conference on Artificial Intelligence had close to 6,000 attendees. Over half of them were venture capitalists, recruiters and media folk. In 1987, *Fortune* magazine—hardly the place for cutting-edge computer research—praised the arrivals of "Live Experts on a Floppy Disk." For the first time in AI's history, researchers were getting as rich as the new PC upstart entrepreneurs like Steve Jobs and Bill Gates.

Interestingly, seasoned researchers like Marvin Minsky shied away from this. It would be easy to assume that the old guard of AI would have been eager to cash in after more than a quarter century of hard work. In fact, they were waiting for the other shoe to drop. It didn't take long. As with the speculative dot-com bubble of the late 1990s, exponents tended to overstate the abilities of expert systems to a dangerous degree. One textbook invoked the "phone-call rule" suggesting that "any problem that can be and frequently is solved by your in-house expert in a ten to thirty-minute phone call can be automated as an expert system." The underlying concept of expert systems was solid, but they had problems. They were expensive, required constant updating and—counterintuitively—could become *less* accurate the *more* rules were incorporated. "As rule sets become larger, undesirable interactions between rules become more common, and practitioners found that the certainty factors of many other rules had to be 'tweaked' when more rules were added," Stuart Russell and Peter Norvig write in the textbook *Artificial Intelligence: A Modern Approach*.

In the fiscal year ending 1987, two of the leading expert system companies—Teknowledge and Intellicorp—lost millions of dollars. Other AI companies fared even worse—filing for bankruptcy, leaving employees and executives out in the cold. After a warm spell, AI's second winter was back.

What's All the Fuss About Now?

AI's following cold snap was worse than its first. Money dried up again. Government grants vanished once more. The budget for AI research from the US Defense Advanced Research Projects Agency, DARPA (the new name for ARPA from 1972), declined by a full one-third between 1987 and 1989. Advertising rates fell in specialist Artificial Intelligence magazines. When *Daedalus*, the official journal of the American Academy of Arts and Sciences, dared publish an entire issue on AI in 1988, the philosopher Hilary Putnam was outraged. "What's all the fuss about *now*?" Putnam wrote. "Why a whole issue of *Daedalus*? Why don't we wait until AI achieves something and *then* have an issue?" The backlash was felt throughout the tech world. Membership in the Association for the Advancement of Artificial Intelligence tailed off. By its nadir in 1996, it had plummeted to just 4,000 members worldwide. Short of a miracle, the dream of Artificial Intelligence appeared to be over.

That year, two students at Stanford—one the child of an AI researcher, the other of a mathematician—came up with a clever way to build a smart web catalogue by ranking pages based on the number of incoming links. In 1997, twenty-four-year-old Larry Page and Sergey Brin turned their nifty algorithm into a company, launched from a garage in Menlo Park. To make it the "Worldwide Headquarters" they thought it should be, they kitted it out with a few tables, three chairs, a turquoise shag rug, a folding Ping-Pong table and a few other items. The garage door had to be left open for ventilation.

It must have seemed innocuous at the time, but over the next two decades, Larry Page and Sergey Brin's company would make some of the biggest advances in AI history. These spanned fields including machine translation, pattern recognition, computer vision, autonomous robots and far more, which AI researchers had struggled with for half a century.

Virtually none of it was achieved using Good Old-Fashioned AI.

The company's name, of course, was Google.

2

Another Way to Build AI

IT IS 2014 and, in the Google-owned London offices of an AI company called DeepMind, a computer whiles away the hours by playing an old Atari 2600 video game called *Breakout*. The game was designed in the early 1970s by two young men named Steve Jobs and Steve Wozniak, who later went on to start a company called Apple. *Breakout* is essentially a variation on the bat-and-ball tennis game *Pong*, except that instead of hitting the square "ball" across the screen to another player, you fire it at a wall of bricks which smash on impact. The goal is to destroy all of the bricks.

As we saw in the previous chapter, there is nothing at all unusual about AI playing games. Alan Turing wrote the world's first chess program as far back as 1947, although computers were not yet powerful enough to run it at the time. Today video games feature plenty of non-player-controlled characters, which are programmed with simple rules that combine to give rise to complex behaviors. So what is so special about DeepMind's game playing?

There are two answers to this question. The first is that it gets better as it plays. Like seeing your child grow up, the change is barely noticeable if you watch the computer constantly. Drop in every fifty or so games, however, and the effect is startling. At first, DeepMind's AI is crushingly awful at *Breakout*. It misses easy shots and seems baffled about what's going on: like handing a PS4 controller to your ninety-year-old great-aunt and expecting her to immediately understand what she's meant to do. Points are scored occasionally, but even the most optimistic of onlookers would be hard-pressed to call them anything more than accidents.

But by the 200th game, things are different. Now the paddle skips skittishly back and forth across the screen: scoring easily, if not consistently. Another few hundred games and DeepMind's AI is the equivalent of Luke Skywalker at the end of *Star Wars: A New Hope* or Neo from *The Matrix*—effortlessly batting the square ball back and forth with a lazy ease. All signs of extraneous movement are gone, and a clear strategy has emerged.

The second reason DeepMind's AI is so significant is because it does not require masses of human-led training. The central tenet of Good Old-Fashioned AI is that rules had to be pre-loaded into the system, like a teacher preparing a child for an exam by having them learn every answer in order. DeepMind, instead, learns on its own. It also does this without having to even be told what it is supposed to be doing. All it has access to are the 30,000 pixels that make up each frame of *Breakout* and the on-screen player score. With nothing more than the instruction to maximize its score, it picks up the "rules" by which the game is played and then hones the strategies needed to perfect them.

Nor is *Breakout* the only game it can play. DeepMind's AI started out playing *Space Invaders*, and has also learned forty-eight additional titles with the sparsest of information. These include boxing simulators, martial-arts titles and even 3-D racing games. There is still a distance to go until it moves beyond the micro-world of a retro video game, but it remains an astonishing achievement that hints at the next step in AI's life cycle. That step? According to DeepMind's own mission statement, it is no less than to "solve intelligence."

The Importance of Learning

Learning is a profoundly important part of what makes us human. It is also something Good Old-Fashioned AI struggled with. The kind of systems described in chapter one were capable of learning only insofar as they could follow rules that had been extracted from the knowledge of "knowledge engineers" and then codified into system architecture. It was a top-down imagining of knowledge and carried an implicit assumption that this was something which could not be automatically learned by a machine. Instead, knowledge had to be programmed, one piece at a time. In many scenarios this could be done well enough to perform limited tasks at acceptable level. The problem came with scaling the solutions. Like any bureaucracy, they become big, unwieldy, slow and expensive.

This presented an obvious problem. As Pedro Domingos, professor of computer science at the University of Washington, said: "If a robot had all the same capabilities as a human except learning, the human would soon leave it in the dust." But right from the start, there has been a parallel vision for Artificial Intelligence, which is now triggering many of the advances we are seeing in the field. Rather than conceptualizing a mind, this school of AI is rooted on modeling the brain inside a computer. Instead of believing that logical reasoning is the best (and perhaps the only) way to achieve true knowledge, it takes an empirical approach rooted in observation and experimentation. And instead of being the work of knowledge engineers, it is the province of a group of computer scientists called machine learners.

This school of AI is dominated by probabilistic models pioneered by statisticians, neuroscientists and theoretical physicists. Much of it is based around what are called "neural networks"—or "neural nets" in computing slang—which function as vast computational approximation of the human brain. In the brain, information is represented by the electrical firing patterns of neurons. There are approximately 100 billion neurons in the human brain, broadly equal to the number of stars that exist in the Milky Way galaxy. Memories are formed through the strengthening of these different neurons firing together: a process called "long-term potentiation." Although we have yet to build a neural network with close to the complexity of the human brain (more on that in a later chapter), computational neural networks borrow the same metaphor for laying down memories and, as a result, learning. The primary difference is that, where long-term potentiation in the brain is a biochemical process, in neural networks learning takes place by modifying its own code to find the link between input and output—or cause and effect—in situations where the relationship is complex or unclear.

Despite their status in the AI community today, for many years neural networks were largely ignored; they were viewed as the maligned stepbrother of true Artificial Intelligence. As David Ackley, a prominent researcher who broke into the field in the 1980s, told me: "When we got into neural nets, it was not considered Artificial

Intelligence. We were rejects from Artificial Intelligence. Artificial Intelligence was symbolic. It was [about] production systems, it was [about] expert systems, and so on. When I got to graduate school at Carnegie Mellon, I was already tired of symbolic, traditional computer-oriented stuff . . . It seemed to me that there had been far too much focus on reasoning, and not enough focus on judgment."

A generation of AI researchers thought like Ackley, and have helped statistical tools all but replace Good Old-Fashioned AI in the mainstream consciousness. In doing so, neural nets have helped achieve things previous generations of researchers could only dream of: whether it be building machines that can learn to play video games, understand speech, recognize individual faces within photographs, or even drive cars on the road more safely than a human driver.

We'll cover some of these applications in this chapter. However, before we get there, we must first go back to the past, and meet a man by the name of Santiago Ramón y Cajal.

The Father of Neuroscience

Santiago Ramón y Cajal was a nineteenth-century Spanish pathologist, often considered the father of modern neuroscience. It was Ramón y Cajal who carried out one of the first detailed examinations of the human brain. Working at the University of Barcelona in 1887, he found that it was possible to use potassium dichromate and silver nitrate to stain neurons a dark color, leaving the surrounding cells transparent. As he later recalled, the stained nerve cells appeared "colored brownish-black even to their finest branchlets, standing out with unsurpassable clarity upon a transparent yellow background. All was sharp as a sketch with Chinese ink." This technique of staining nerve cells meant that Ramón y Cajal was able to make extensive studies of the brain, something which had been previously impossible using contemporary microscopes. In doing so, he was able to prove for the first time that neurons are the building blocks from which the central nervous system is comprised.

In 1943, nine years after Ramón y Cajal's death, a pair of AI researchers, Warren McCulloch and Walter Pitts, created the first formal model of a neuron in an influential paper with the unwieldy title, "A Logical Calculus of the Ideas Immanent in Nervous Activity."

McCulloch and Pitts were an unusual pairing. Warren McCulloch was born in 1898, entering a family of lawyers, engineers, doctors and theologians. He grew up in Orange, New Jersey, which was then known as the hat-making capital of America. After changing his mind on a life in the ministry, McCulloch studied philosophy and psychology at Yale, where he developed an interest in neurophysiology, which is the study of the nervous system.

Pitts was twenty-five years younger than McCulloch. Born in 1923 into a working-class family, he was an unlikely child prodigy. At the age of thirteen, Pitts ran away from home to escape his abusive father and, for a while, lived rough on the streets. One day he was chased by some local bullies and hid in a public library. According to legend, Pitts spent the next week hungrily devouring the three-volume math textbook *Principia Mathematica*. After he had finished, Pitts decided to write to one of the books' authors, Bertrand Russell, pointing out what he perceived as the fundamental errors in the series' first volume. Russell was impressed and went so far as to invite Pitts to study at Cambridge in the United Kingdom, although Pitts was unable to take him up on his offer. In his late teens, Pitts became fascinated by the work of a Russian mathematical physicist named Nicolas Rashevsky, whose work focused on the field of mathematical biophysics. It was in this capacity that Walter Pitts met and eventually began working with Warren McCulloch.

In collaboration, McCulloch and Pitts came up with a simplified model of a functioning neuron replicated inside a machine. In their 1943 paper, they claimed that a neuron was, at its root, a "logic unit." They also demonstrated that a network made up of such units could be made to perform every possible computational operation.

The Neurons That Fire Together, Wire Together

McCulloch and Pitts' work was a crucially important development, but it also had a severe limitation: it was unable to learn. This problem was theoretically solved six years later when the Canadian psychologist Donald Hebb wrote the 1949 book *The Organization of Behavior*. Hebb argued that neural pathways in the brain are strengthened every time they are used, which explains the way in which humans learn. "When an axon of cell A is near enough to excite a cell B and repeatedly or persistently takes part in firing it, some growth process or metabolic change takes place in one or both cells such that A's efficiency, as one of the cells firing B, is increased," he wrote. To put this a bit more simply, Hebb was referring to the idea that when two neurons in the brain fire simultaneously, the connection between them is enhanced. This is sometimes remembered with the rhyme, "The neurons that fire together, wire together."

It took another decade until Hebb's ideas found their way into computer research, thanks to a man named Frank Rosenblatt. Rosenblatt is an intriguing figure in computing history: a polymath and true Renaissance man who appears to have been an expert in everything from music and astronomy to mathematics and computing. As it happened, he had been a classmate of Marvin Minsky, who we met in the last chapter, at the Bronx High School of Science during the early 1940s. However, he had remained on the fringes of mainstream AI research. While Minsky and John McCarthy were organizing the Dartmouth conference, Rosenblatt had earned his PhD in experimental psychology from Cornell University, during which time he had become enamored with the subject of neural networks. Rosenblatt referred to neural networks as "perceptrons" and began working to prove that they could function as effective models for human learning, memory and cognition.

Some of Rosenblatt's first attempts at building a perceptron took place at the Cornell Aeronautical Laboratory in Buffalo, New York. There he created a project named PARA, standing for "Perceiving and Recognizing Automation." His perceptrons built on the neuron model proposed by McCulloch and Pitts, based around neural networks that learn through trial and error. Each neuron had an input, an output and an individual set of "weights." At the start, the connections between what are known as "features" and neurons are assigned random weights. Depending on what the network is shown, the neurons then fire or don't fire. After a while, it develops the ability to classify everything it sees into the category of either "X" or "Not X."

Because computers were so slow at the time, Rosenblatt built his perceptrons as physical pieces of hardware rather than software. Weights were created using variable resistors of the kind used in light dimmers, while the learning process was accomplished with electric motors and resistors. The ensuing demonstration, combined with Rosenblatt's extravagant claims about the possibilities of perceptrons, was enough to get people excited, however. In a strikingly prescient 1958 article, marred by the hyperbolic title "Human Brains Replaced?," a writer for *Science* magazine gushed: "Perceptrons may eventually be able to learn, make decisions, and translate languages." A *New Yorker* article meanwhile quoted Rosenblatt as saying perceptrons should prove capable of telling "the difference between a dog and a cat" using computer vision.

In 1960, Rosenblatt oversaw the creation of an "alpha-perceptron" computer called the MARK I, for which he received sponsorship from the Information Systems Branch of the Office of Naval Research. It became one of the first computers in history to be able to acquire new skills through trial and error. The *New York Times* hailed it as the "New Navy Device [That] Learns By Doing."

The Problem with Perceptrons

Sadly, not long after this, work with perceptrons suffered two serious setbacks. The first was technical . . . and a little personal. Perceptrons had proven capable of simple learning tasks such as recognizing speech sounds or identifying printed letters. However, they also succeeded at generating attention and funding far beyond their modest levels of success. This caused friction in the AI community. One outspoken critic was Marvin Minsky. Minsky had actually studied neural networks for his PhD, but he had become disillusioned with the field. From the late 1950s onward, Rosenblatt debated Minsky at various scientific conferences about the usefulness of brain-inspired computation. Rosenblatt, typically, made enormous claims for his technology, suggesting that perceptrons could carry out virtually any learning task. Minsky argued the opposite. This stalemate continued until 1969, when Minsky coauthored a devastating book with fellow researcher Seymour Papert, attacking everything that perceptrons were supposedly not capable of. Minsky and Papert concluded that the technology was "without scientific value." Funding for neural networks crashed almost overnight.

The second setback was altogether more tragic. Two years after Minsky and Papert's book *Perceptrons* was published, Frank Rosenblatt went on a Sunday boating trip in Chesapeake Bay, the largest estuary in the United States. It was his forty-third birthday. An accident took place and Rosenblatt was killed. In a touching tribute paid to him by the faculty at Cornell, his colleagues wrote, "We have lost, in his passing, one of the most selfless and sympathetic colleagues, whose good humor and brilliant mind left a deep impression on us all." Perceptrons had also lost their staunchest supporter.

Brain-inspired neural nets appeared to be dead for the next decade. Seymour Papert later summarized the clash between Good Old-Fashioned AI and perceptrons by relating them to a fairy tale.

Once upon a time, two daughter sciences were born to the new science of cybernetics. One sister was natural, with features inherited from the study of the brain, from the way nature does things. The other was artificial, related from the beginning to the use of computers. Each of the sister sciences tried to build models of intelligence, but from very different materials. The natural sister built models (called neural networks) out of mathematically purified neurons. The artificial sister built her models out of computer programs.

Borrowing from the tale of Snow White, Papert likened the impact of his and Minsky's assault on perceptrons to the huntsman sent out into the woods to kill the titular heroine. As in the fairy tale, Papert and Minsky returned home to their master (in this case, "Lord DARPA") with the perceptrons' "heart as proof of the dead." Just as in Snow White, however, Papert notes that Snow White was not dead at all. "What Minsky and Papert had shown the world as proof was not the heart of the princess, it was the heart of a pig."

Comically overwrought it might have been, but Papert was correct when he acknowledged that neural nets had survived the onslaught from himself and Minsky. In fact, by the time Papert wrote this in the late 1980s, neural networks were back in full swing.

The Rise of Hopfield Nets

Contrary to Minsky and Papert's assertions, researchers working with neural networks had for some years believed that neural nets could take on new abilities—and solve the problems with Rosenblatt's perceptrons—only if extra "hidden" layers of neurons could be placed between the network's input and output. Unfortunately, no one knew how to train these multilayer networks. The man responsible for suggesting how this might be done was a renowned physicist named John Hopfield.

Hopfield had no great interest in what was then the mainstream form of Artificial Intelligence. "I never actually dug deeply into what was going on in AI," he says. "It was so incompetent when faced with any real-world problems, I didn't feel I needed to learn about it." For years, however, he had been searching for what he refers to as the "problem of a lifetime" to sink his teeth into. An interest in the brain made him consider everything from primate neuroanatomy and insect flight behavior, to learning in the rat hippocampus or curing Alzheimer's disease. For a while, Hopfield was fascinated by cellular automata and the prospect of robots that could build copies of themselves. However, after months of research, it led him to a dead end.

"It is surprisingly difficult to give up on a wrong idea that has been nurtured for a year," Hopfield says. But the idea of creating a model of life inside a computer stayed with him. He was fascinated by the idea of using a network to accomplish a task which the brain does rapidly and easily, but which computers were incapable of. He settled on the idea of associative memory, which describes the way that the brain is capable of working reciprocally—meaning that seeing a person reminds you of their name, or hearing their name reminds you of what they look like. Thinking about the mathematics behind associative memory somehow reminded Hopfield of the mathematics of what are called "spin" systems, describing the complex forms of magnetism in solids. A light went on in his head. "Suddenly there was a connection between neurobiology and physics systems I understood," Hopfield recalls. "A month later I was writing a paper."

The result of this 1982 paper was the creation of a whole new type of neural network. Hopfield networks were more complex than the single layer of simulated neurons found in Rosenblatt's perceptrons. His insight helped revive interest in neural networks and made him an unlikely hero in the process. At Caltech, a group of interested followers began meeting under the name "Hop-Fest." Hopfield's discovery helped attract some of the world's greatest theoretical physicists to come and work with neural nets. For the first time in years, researchers in the field began to get excited.

Not that things were necessarily easy. As we saw in chapter one, the early years of the 1980s were dominated by "expert systems," now with more money than ever behind them. Although they would stumble later on, at the time they appeared to be too big to fail. "We felt like little furry mammals at the time of the dinosaurs," recalls Terry Sejnowski, now one of the world's leading neural network experts, who was then Hopfield's doctoral student at Princeton. "We were screwing around under the legs of these huge behemoths who were out there with multimillion-dollar machines and enormous budgets. Absolutely everyone was focused on computational logic at the time, but it was clear to us that they were ignoring the really difficult problems that we knew were crucial for driving AI forward."

Fortunately, neural nets were picking up a number of enthusiastic young researchers. These included the likes of David Rumelhart and James McClelland, two cognitive scientists at the University of California San Diego, who formed an artificial neural network group that became incredibly influential in its own right.

There was also a man named Geoff Hinton.

The Patron Saint of Neural Networks

Born in 1947, Geoff Hinton is the one of the most important figures in modern neural networks. An unassuming British computer scientist, Hinton has influenced the development of his chosen field on a level few others can approach. He comes from a long line of impressive mathematical thinkers: his great-great-grandfather is the famous logician George Boole, whose Boolean algebra laid the foundations for modern computer science. Another relative was Charles Howard Hinton, a mathematician noted for his ideas about four-dimensional space, who is mentioned twice in Aleister Crowley's novel *Moonchild*.

"I was always interested in how people thought and how the brain worked," Hinton says. At school, a friend convinced him that the brain stores memories in much the same way that a 3-D holographic image stores light information. To create a hologram, people bounce multiple beams of light off an object and then record these bits of information in a giant database. The brain does the same, only with networks of neurons instead of beams of light. This observation led Hinton to study physiology and psychology at Cambridge and then Artificial Intelligence at the University of Edinburgh in Scotland. He arrived in the chilly city of Edinburgh in the mid-1970s, appropriately enough just at the time the first AI winter was setting in. Despite the blow that Good Old-Fashioned AI had just suffered, Hinton's doctoral supervisor desperately tried to steer him away from neural networks. "He kept trying to get me to give up on them and switch to symbolic AI," he says. "We kept making deals where I would get to do neural nets for a little bit longer."

Hinton didn't get much support elsewhere. His fellow students thought he was crazy to be studying neural networks after Minsky and Papert had so totally decimated the field. While Hinton was at Edinburgh, one of the first Artificial Intelligence textbooks was published, written by the influential AI researcher Patrick Winston, a graduate student of Minsky. The entry on neural networks reads as follows:

> Many ancient Greeks supported Socrates' opinion that deep, inexplicable thoughts came from the gods. Today's equivalent to those gods is the erratic, even probabilistic neuron. It is more likely that increased randomness of neural behavior is the problem of the epileptic and the drunk, not the advantage of the brilliant.

Winston might have been overly dismissive (later revisions of the textbook toned down the assault), but he wasn't entirely wrong about the almost religious faith needed to believe in neural nets at the time. Hinton was comforted by the knowledge that the brain must work somehow, and that it quite clearly wasn't explainable using symbolic AI. "Most of the common-sense reasoning we do is done intuitively and by analogy," he says. "It doesn't involve a big sequence of conscious operations." This, Hinton felt, was the "cheat" of Good Old-Fashioned AI: that everything is a series of basic rules and conscious reasoning. To the symbolic AI researchers, if there is a part of consciousness we do not understand, it is only because we haven't worked out the reasoning behind it.

After Hinton graduated, he briefly carried out postdoctoral work in Sussex, before a job offer in the United States materialized. Hinton packed up his things and moved to the University of California San Diego and, soon after, to Carnegie Mellon University. For the next several years, he was responsible for a slew of groundbreaking advances in neural networks, which continue to reverberate in AI labs around the world today.

Perhaps the most significant of these was helping another researcher, David Rumelhart, rediscover the "back-propagation" procedure, arguably the most important algorithm in neural networks, and then producing the first convincing demonstration that backpropagation allowed neural networks to create their own internal representations. "Backprop" allows a neural network to adjust its hidden layers in the event that the output it comes up with does not match the one its creator is hoping for. When this happens, the network creates an "error signal," which is passed backward through the network to the input nodes. As the error is passed from layer to layer, the network's weights are changed so that the error is minimized. Imagine, for example, that a neural net is trained to recognize images. If it analyzes a picture of a dog, but mistakenly concludes that it is looking at a picture of a cat, backprop lets it go back through the previous layers of the network, with each layer modifying the weights on its incoming connections slightly so that the next time around it gets the answer correct.

A classic illustration of backprop in action was a project called NETtalk, an impressive demo created in the 1980s. Cocreator Terry Sejnowski describes NETtalk as a "summer project" designed to see whether a computer could learn to read aloud from written text. The challenge with this is that language is not at all straightforward. At the start of the project, Sejnowski went to the library and took out a book on phonology by Noam Chomsky and Morris Halle, called *The Sound Pattern of English*. "It was filled with rules about things like how you pronounce the letter 'e' when it's at the end of a word in such-and-such a context," Sejnowski says. "There were exceptions; then there were exceptions to the exceptions. English was just a huge mess of complex interactions. It turns out that we had chosen one of the worst languages in the world when it comes to irregularities."

Plugging each of these separate examples into an expert system is how the Good Old-Fashioned AI community would have attempted to carry out the task. Sejnowski and his colleague, a language researcher named Charles Rosenberg, decided instead to create a 300-neuron neural network to achieve the goal. Hinton, who was visiting the lab at the time, suggested that the pair start their project by training it with a children's book, featuring a very limited vocabulary. At first, it was a laborious process. The computer moved through each word one letter at a time, and for each letter the pair had to attach the correct phoneme. For instance, the letter "e" is pronounced differently depending on whether you're reading "shed," "pretty," "anthem," "café" or "sergeant." Each time Sejnowski and Rosenberg offered clarification, their neural network silently adjusted the weights on each connection. The biggest challenge was getting the machine to correctly pronounce the middle part of each word. To do this, the neural net had to use the context provided by the letters to the left and right of the center.

By the end of the day, NETtalk had mastered the entire 100-word book. They were thrilled. Next, they set NETtalk to take on a 20,000-word Webster's dictionary. Fortunately this had all of the phonemes already marked. They set it going in the afternoon and went home for the day. When they came into the office the following morning, it had completely mastered it.

The final piece of training data was a book featuring a tran-

scription of children talking, along with a list of the actual phonemes spoken by the child, written down by a linguist. This meant
that Sejnowski and Rosenberg were able to use the first transcript
for the input layer and the second phoneme transcript for the output. By using backprop, NETtalk was able to learn exactly how to
speak like a real kid. A recording of NETtalk in action shows the
rapid progress the system made. At the start of training, it can
only distinguish between vowels and consonants. The noise it produces sounds like vocal exercises a singer might perform to warm
up his or her voice. After training on 1,000 words, NETtalk's
speech became far more recognizably human. "We were absolutely
amazed," Sejnowski says. "Not least because computers at the time
had less computing power than your watch does today."

The Connectionists

Aided by the work of Geoff Hinton and others, the field of neural
nets boomed. In the grand tradition of each successive generation
renaming themselves, the new researchers described themselves as
"connectionists," since they were interested in replicating the neural connections in the brain. By 1991, there were 10,000 active con-
nectionist researchers in the United States alone.

Suddenly, groundbreaking demonstrations were everywhere.
For instance, neural networks were discovered to be particularly
good at predicting the stock market. In most cases, investment
firms trained separate networks for different stocks, with human
traders then deciding which to invest in. However, some went further and gave the networks themselves the autonomous power to
buy and sell. Not coincidentally, the finance sector quickly joined

the video game business as an industry ready to throw money at AI researchers. The age of algorithmic trading had begun.

Another eye-catching application of neural nets during this time was the invention of the self-driving car. Autonomous vehicles had been a long-time dream of technologists. In 1925, the inventor Francis Houdina demonstrated a radio-controlled car, which he drove through the streets of Manhattan without anyone at the steering wheel. Later, autonomous vehicle tests used guidewires and on-board sensors to follow painted white lines on the road or seek out the alternating current of buried cables. In 1969, John McCarthy came closest to describing modern self-driving vehicles when he wrote an essay with the provocative title "Computer-Controlled Cars." What McCarthy was proposing, he wrote, was essentially "an automatic chauffeur." His project called for a computer capable of navigating a public road, equipped only with a "television camera input that uses the same visual input available to the human driver." McCarthy imagined users being able to enter a destination using a keyboard, which would prompt the car to immediately drive them there. Additional commands would allow users to change destination, stop at a restroom or restaurant, slow down or speed up in the case of an emergency.

Such projects came to nothing until the early 1990s, when a Carnegie Mellon researcher named Dean Pomerleau wrote an exciting PhD thesis, describing how back-propagation could be applied to a self-driving vehicle. Pomerleau's neural network—which he called Autonomous Land Vehicle in a Neural Network, or ALVINN—took raw images from the road as its input and output steering controls in real time. A number of other Good Old-Fashioned AI PhD candidates were working on similar self-driving projects at the time. These non-neural-net approaches focused on logically segmenting each image into categories like "road" and "non-road" through careful pixel analysis. As with many classic AI vision problems, however, the computer had difficulty parsing information as unstructured as real roads. Given that a self-driving car would be moving at dangerous speeds while relying on this technology, the potential for disaster was high. "They would sometimes classify the shadow from a tree, or even the tree itself, as a road due to the straight converging lines," Pomerleau recalls. "So they'd head directly for the tree rather than try to avoid it."

To train ALVINN, a human driver simply had to drive along a stretch of road. "It would take about two or three minutes of the human driver driving, while the ALVINN system learned and updated the weights of the backprop net," Pomerleau says. "By the end of that time, the driver could let go of the wheel and the system would continue driving on an entirely new stretch of road." Pomerleau's creation only focused on steering and was unable to manage speed control or obstacle avoidance, both of which had to be carried out by the human driver. Nonetheless it was successful enough that, in 1995, an advanced version of ALVINN, called RALPH (for Rapidly Adapting Lateral Position Handler), was installed in a Pontiac Transport minivan which had been reclaimed from a junk heap. Kitting it out with a computer, 640 x 480 color camera, GPS receiver and fiber-optic gyro, Pomerleau and a fellow researcher named Todd Jochem set out to drive across the United States. In a reference to the 1986 charity event "Hands Across America," they named the trip "No Hands Across America." To help pay for hotel rooms and food, the pair sold $10 T-shirts along the way. In the end, the car drove 2,797 miles coast to coast from Pittsburgh, Pennsylvania, to San Diego, California—including a crossing of the Hoover Dam carried out autonomously. In one memorable highlight, a *Businessweek* reporter who was covering the event was pulled over by a Kansas State Trooper. Pomerleau and Jochem sailed by in their self-driving car, hands exaggeratedly off the steering wheel.

It would be another fifteen years, until October 2010, before Google announced its own self-driving car initiative. However, thanks to his groundbreaking work in neural nets, Dean Pomerleau had proved his point.

Welcome to Deep Learning

The next significant advance for neural networks took place in the mid-2000s. In 2005, Geoff Hinton was working at the University of Toronto, having recently returned from setting up the Gatsby Computational Neuroscience Unit at University College London. By this time it was clear that the Internet was helping to generate enormous data sets that would have been unimaginable even a decade before. If previously researchers had had the problem of not enough data to properly train their networks, the rise of the Internet changed this in a profound way. Today, research firms such as International Data Corporation estimate the amount of currently existing online data stands at around 4.4 zettabytes, or 4.4 trillion gigabytes. As journalist Steve Lohr points out in his highly entertaining book *Data-Ism*, should this amount of data ever find its way onto Apple's super-slimline iPad Air tablets, the ensuing stack would reach two-thirds of the way to the moon.

However, much as the Earth is covered with water, but not all of it is immediately drinkable, the problem with a lot of this data is

that it is unlabeled. When data sets were smaller, far more attention was paid to keeping them labeled correctly, making it far
more useful for training networks. As the amount of data ex-
ploded, this became impossible. For instance, in March 2013,
Flickr had a total of 87 million registered users uploading in excess
of 3.5 million new images daily. In theory, that is great news for
people wanting to build a neural network for recognizing images,
but it raises challenges, too. As we've seen, the simplest way to
train a neural network is to show it a large number of images and
then to tell it what each of the images is. By labeling the images,
the trainer provides both an input (the image) and output (the de-
scription). They can then back-propagate to correct any mistakes.
This is what is known as "supervised learning." But with so many
unlabeled or incorrectly labeled images in circulation, how is the
computer to learn?

Fortunately, Geoff Hinton triggered a revolution in what is
called "unsupervised learning," in which no labels at all are provided to the computer. All the machine has access to is an input,
with no explanation at all of what it is looking at. At first, it sounds
like it should be impossible for a machine to learn in this way.
Nobody, not even the smartest neural network, can be expected
to learn what something is if they are never explicitly told. In fact,
what Hinton discovered was that unsupervised learning could be
used to train up layers of features, one layer at a time. This was the
catalyst in the field of "deep learning," currently the hottest area
in AI.

You can think of a deep learning network a bit like a factory
line. After the raw materials are input, they are passed down the
conveyor belt, with each subsequent stop or layer extracting a different set of high-level features. To continue the example of an im-

age recognition network, the first layer may be used to analyze pixel brightness. The next layer then identifies any edges that exist in the image, based on lines of similar pixels. After this, another layer recognizes textures and shapes, and so on. The hope is that by the time the fourth or fifth layer is reached, the deep learning net will have created complex feature detectors. At this stage, the deep learning net knows that four wheels, a windshield and an exhaust pipe are commonly found together, or that the same is true of a pair of eyes, a nose and a mouth. It simply doesn't know what a car or a human face is. Many of the features it recognizes may not be relevant to the task at hand, but some of them will be highly relevant.

"The idea was that you train up these feature detectors, one layer at a time, with each layer trying to find structural patterns in the layer below. Once you've done that, you then stick the labels on top and apply back-propagation to fine-tune everything," Hinton explains. The result sent a shock wave through the AI community. "There was also some nice maths in there," Hinton recalls, wryly. "That always impresses people."

Word quickly got out about deep learning. Two members of Hinton's lab, George Dahl and Abdel-rahman Mohamed, quickly demonstrated that it worked just as well for speech recognition as it did for image recognition. In 2009, the pair pitted their newly created speech recognition neural network up against the then–industry standard tools, which had been worked on for the past three decades. The deep learning net won. At this point, major companies began to take an interest. One of these was Google. In 2011, a PhD student of Hinton's named Navdeep Jaitly was asked to tinker with Google's speech recognition algorithms. He took one look at them and suggested gutting the entire system and replacing it with a deep neural network. Despite being initially skeptical, Jaitly's boss agreed to let him try. The program outperformed the system Google had been fine-tuning for years. In 2012, Google incorporated deep learning speech recognition into its Android mobile platform. Instantly the error rate dropped 25 percent from where it had been previously.

That summer, Hinton received a long-overdue call from Google. The search giant invited him to spend the summer working at its enormous campus in Mountain View, California. Although he was sixty-four years old at the time, Google hilariously classified Hinton as an "intern," because its strict adherence to policy dictated that its title of "visiting scientist" could only be bestowed on someone staying longer than a few months. Gamely, Hinton agreed to join the intern group, which consisted mainly of students in their early twenties. He even donned the beanie hat with a propellor on top that is given to all new Google interns, who are referred to as "Nooglers," at the company's TGIF drinks bash. "I must have been the oldest intern they ever had," Hinton says. Commenting at the time, he jokingly suggested that his much younger colleagues—unaware of who he was—viewed him as a "geriatric imbecile."

Hinton's job at Google involved advising on other potential applications for deep learning. The summer went well, and the following year Google hired him for real. The company also hired two more of his graduate students, with whom he had created a startup called DNNresearch. In a statement, Hinton wrote that, "I'll remain part-time at the University of Toronto, where I still have a lot of excellent graduate students, but at Google I will get to see what we can do with very large-scale computation."

After thirty years' plowing an often lonely furrow with neural networks, Geoff Hinton was finally playing a key role in the biggest AI company in the world.

The New AI Mainstream

Today, deep learning neural nets have become the mainstream in Artificial Intelligence, reaffirming ideas that trace all the way back to McCulloch and Pitts. While it is still only an approximation of how the brain actually works (we will discuss the existence of more biofidelic models of the brain in a later chapter), what is impressive is the broad range of problems neural nets can be applied to. Unlike Good Old-Fashioned AI, which worked perfectly until it discovered that the world didn't match up to its perfect model, neural nets perform as well with exceptions to rules as they do with regularities. As NETtalk showed in the 1980s, this makes them perfect for dealing with tricky areas like language. They're also good at what are called "distributed representations," meaning that they possess the ability to model two seemingly separate areas (such as language and image) in the same representational space. In essence, this means neural networks can think using analogies, which is something that could never be said for classic AI.

"A lot of the things that are now working are because people are using neural nets," observes Geoff Hinton. "The rule of thumb is that if there's a task you want to do and you know it involves huge amounts of knowledge, that means that if you're going to learn to do it you need huge numbers of parameters. If that's the case, then deep learning is the way to do that."

Impressive applications are everywhere. In 2011, the summer

before Hinton joined Google, Google engineers Jeff Dean and Greg Corrado and Stanford computer scientist Andrew Ng launched what is known as the Google Brain project. Housed at Google's semi-secret research laboratory, Google X, Google Brain used a deep learning net to recognize high-level concepts such as cats by analyzing still images from YouTube videos—without ever having been told what a cat is. (Incidentally, this is virtually the exact same feat Frank Rosenblatt told the *New Yorker* that neural networks would one day perform, half a century earlier.)

Having a computer that knows what a cat is may not sound like a particularly useful achievement, but the ability to use deep learning for computer vision has a host of real-world uses. One startup called Dextro is using deep learning to create better tools for online video searches. Instead of relying on keyword tags, Dextro's neural net scans through live videos, analyzing both audio and image. Ask it about David Cameron, for example, and it will bring up not just Conservative Party videos, but also video in which the UK prime minister is only mentioned in passing.

Facebook, meanwhile, uses deep learning to automatically tag images. In June 2014, the social network published a paper describing what it refers to as its "DeepFace" facial recognition technology. Thanks to deep learning, Facebook's algorithms have proven almost as accurate as the human brain when it comes to looking at two photos and saying whether they show the same person, regardless of whether different lighting or camera angles are used. Facebook is also using deep learning to create technology able to describe images to blind users—such as verbalizing the fact that an image shows a particular friend riding a bicycle through the English countryside on a summer's day.

Other projects combine deep learning with robotics. One

group of researchers from the University of Maryland has taught a robot how to cook a simple meal by simply showing it "how-to" cooking videos available on YouTube. Without any direct human input, the robot can then replicate tasks shown in the video with a high degree of accuracy, so long as it is provided with the right utensils to do so. Long-term, the plan is that similar robotic deep learning could be used in areas like military repair.

Deep learning nets have additionally proven essential in translation tasks. In December 2012, Microsoft's research chief Rick Rashid demonstrated a spectacular live English-to-Chinese voice recognition and translation system. Like the *Star Trek* dream of the universal translator, this technology means that in the near future it will be possible to order dinner in a French restaurant, give detailed directions to a taxi driver in Russia, or discuss a potential business deal in Japan without speaking a word of French, Russian or Japanese. Even more impressively, the deep learning system carries out the translation in the user's own voice by breaking their speech into its elemental phonemes and then reassembling them to make up the sounds of whichever language is required. As Microsoft explains, "Your tablet or smartphone will do the heavy lifting of understanding what you're saying in English, translating it into your listeners' tongue, and speaking it in your voice with the pronunciation, tones, and inflections of a native speaker."

Interestingly, while there have been some tweaks to the underlying technology, many of today's big advances come back to the same back-propagation algorithm that David Rumelhart and Geoff Hinton rediscovered in the 1980s. What has changed is the amount of computing power, which in turn means bigger neural networks, with more hidden layers. The Google Brain project alone linked together a massive 16,000 computer processors to create a vast artificial brain with more than 1 billion connections. There has also been a massive increase in the size of the training data sets available. Unlike the comparatively small amounts of data used in previous decades, today there is a veritable flood of usable information to teach neural networks to think. For instance, Facebook's facial recognition system was trained by analyzing around 7.4 million images, taken from Facebook's 1.23 billion active users.

While neural networks are not the only form of AI being practiced today (we'll discuss other approaches in later chapters), their ascendancy has taken AI to its highest levels of success to date. Unlike classic AI, they're no longer confined to simple lab environments.

In fact, the next chapter will discuss what happens when Artificial Intelligence leaves the confines of what we traditionally think of as computer systems and follows us out into the real world.

3

Intelligence Is All Around Us

IN 1998, THE year in which Apple unveiled its bulbous iMac computer, Harry Potter was introduced to the world and the first portable MP3 player went on sale, a forty-four-year-old professor from the University of Cybernetics in Reading underwent an unusual operation. The aim of Professor Kevin Warwick's elective surgery was to have a silicon chip encased in a glass tube inserted under the skin on his left arm. Once implanted, this radio-frequency identification device (RFID) chip then sent radio signals, via antennae located around his laboratory, to a central computer able to control Warwick's immediate surroundings. "At the main entrance [of my lab], a voice box operated by the computer said 'Hello' when I entered," Kevin Warwick later wrote of his experience. "The computer detected my progress through the building, opening the door to my lab for me as I approached it and switching on the lights. For the nine days the implant was in place, I performed seemingly magical acts simply by walking in a particular direction."

Almost twenty years after Warwick's experiment, it remains shocking, headline-provoking stuff—which, from looking at other stories from Warwick's career, was exactly the point. However, in the decades that have passed, our sense of surprise has likely changed somewhat. While it is still easy to balk at the reason someone would willingly undergo such an invasive procedure, the question of *why* someone should want such a thing has receded into the background. As I write this, I have on my wrist an Apple Watch. It is the 42-mm stainless steel model with Apple's Milanese Loop band. It cost $747 and can do far, far more than Kevin Warwick ever dreamed his RFID implant would achieve. If I receive a text message or a phone call, or if one of my friends posts a new photo to Instagram, I can view it simply by glancing at my wrist. In supermarkets, I can pay for my groceries by tapping my wrist against the card reader. I can do the same to unlock the door to my room at hundreds of hotels around the world. If I'm out walking, a series of taps and vibrations emitted by my watch tell me which way to turn. One series of taps means turn right. Another series means turn left. A first vibration indicates that I'm on the final leg of my journey, while a second vibration tells me that I've arrived. And all without an invasive surgical procedure, too.

If you're reading this book, the chances are that you're familiar with the term "smart devices." In addition to a growing category of smart watches, which also includes Pebble, Android Wear and a variety of others, there are also smart running shoes, capable of tracking numbers of steps and heartbeats, and communicating your mood using embedded screens to show emoji such as smiley faces and hearts. There are smart fridges, which keep track of their temperature and contents, and let you know when your favorite food is running out or about to go bad. And there are smart security cameras, smart kitchen scales, smart lightbulbs, smart toilets, smart diapers and even smart toothbrushes. The most prominent smart device company, Nest Labs, was acquired by Google in January 2014 for a jaw-dropping $3.2 billion in cash. Founded by former Apple employees Matt Rogers and iPod creator Tony Fadell, it builds a range of connected smart devices, chief among which is a smart thermostat, designed to "learn" its user's habits over time and adjust itself accordingly.

What makes these devices "smart" is a combination of sensors, Artificial Intelligence algorithms and constant Internet connectivity via Wi-Fi. Previously, access to web intelligence was something a person had to "jack into." Today, our online connection is rarely, if ever, broken. In aggregate, these advances make it possible to collect data from users, share it, and help users make sense of what it means. "Data empowers us," says Renee Blodgett, president of marketing and strategy for Kolibree, makers of the world's first connected electric toothbrush. "For the first time, we have data on how we brush our teeth, where we brush our teeth and where we need to improve." Before we had smart toothbrushes (which, for me, would be right now), we had to rely on feedback from our dentist once a year when we have our annual check-up. With a smart toothbrush, those lessons are learned and imparted to us in real time.

Forget Electricity, Here's Cognicity

Right now we are in the "early adopter" stages of what will, its boosters claim, be as big a shift as the arrival of electricity in the late nineteenth and early twentieth centuries. In 1879, the American inventor Thomas Edison was able to produce a reliable, long-lasting electric lightbulb in his laboratory in Menlo Park, California. By the 1930s, this technology was available to 90 percent of people living in US cities, and a growing number of rural areas. At the flick of a switch, electricity gave people control over the light in their homes and workplaces, independent of time of day. It interrupted the regular biological rhythms of life and endowed people with a sovereignty over daylight that allowed them to create their own schedules for both work and play. The accompanying network of wires ushered in a slew of connected devices that created industries and changed lives forever.

A spring 1917 catalogue for Sears (then a fledgling mail-order company) advises the public to "Use Your Electricity for More than Light." They did exactly that. Electric irons, washing machines and electric vacuums made laundry and cleaning easier. Cleanliness rose, and families employed fewer domestic servants as a result of the improved efficiencies. Electric refrigerators replaced iceboxes and made it easier to extend the life of food. For the first time, the climate could be controlled through the use of fans when it was hot, and radiant heaters when it was cool. Electricity brought telephones and radio to the masses, and welcomed in an age of instantaneous personal communications for both news and entertainment. By 1938, the former US president Franklin Roosevelt, speaking in Barnesville, Georgia, proclaimed electricity "a modern necessity of life."

Could we be at the start of a similarly transformative journey for smart devices? Perhaps so. Certainly, the rise of mobile wireless networks means that devices are more portable than ever. The dream of what is sometimes (and quite clumsily) termed the "Internet of Things" is that intelligent hardware will become as much a "modern necessity of life" in the twenty-first century as electricity did 100 years ago. Where once we electrified, now we will cognitize.

Right now, hype is so strong around the field of smart devices that analysts at Ericsson predict that there will be in the region of 50 *billion* smart devices around the world by 2020: a figure that works out as approximately 6.8 per person. "This isn't just evolutionary; this is revolutionary," says Michael Grothaus, a former Apple employee who now runs the startup SITU, making smart scales to quantify your calorie intake. "This is the most exciting thing in tech since the personal computer."

Things That Think

In 1991, researchers in the Trojan Room of the computer science department at Cambridge University came up with a neat idea. Allocated a single communal coffee pot among them, they decided to set up a camera to monitor its levels throughout the day. This camera was programmed to capture one frame per second and encode it as a grayscale JPEG file, before sending it over a nascent version of the World Wide Web. From their respective computers, researchers in the department could then save themselves wasted effort by logging in to the "video" feed to see if there was coffee remaining in the pot.

"Some members of the 'coffee club' lived in other parts of the building and had to navigate several flights of stairs to get to the coffee pot—a trip which often proved fruitless if the all-night hackers of the Trojan Room had got there first," explains computer scientist Quentin Stafford-Fraser, who worked in the department at the time. "This disruption to the progress of computer science research obviously caused us some distress, and so XCoffee was born."

I bring up XCoffee because it illustrates an important point about what we consider to be "smart" technology. XCoffee is often singled out as an early example of the modern trend for smart devices. In some senses this is true. As with many newer smart gadgets, XCoffee was connected to the web, and was therefore part of the so-called "Internet of Things." But to me it is closer to an example of what hardware geeks would call a "hack"—a term that colloquially refers to a clever solution to a tricky problem. The prerequisite of what we would today think of as a smart device (fondly described by MIT's Media Lab as a "thing that thinks") is that it exists as a self-governing feedback loop, capable of operating autonomously without a lot of human intervention. The Internet of Things is not simply about "things" connected to the Internet. The traditional Internet was there to allow humans to carry out tasks, such as searching, downloading music, or reading information. The Internet of Things, on the other hand, is designed for non-human entities to communicate, which is why a growing number of people prefer to talk about M2M communication, meaning "machine-to-machine."

A smart device should be able to sense its environment, which leads to an identification of a particular state, which triggers an assessment, which prompts an action, and so on in a continuous loop. The "smart" parts of a smart device are the bits in between, residing in how the sensed information is processed and used to select an action to take. A truly smart coffee machine wouldn't just alert people that it was empty, but work out when people who use it are likely to be thirsty. At this point it should refill itself and produce cups of coffee to fit the individual requirements of people who use it. Even drone-based desk-to-desk delivery shouldn't be out of the question.*

A Brief History of Cybernetics

Most of the smart devices we will discuss in this chapter incorporate elements of machine learning. However, just as the broader questions surrounding Artificial Intelligence date back hundreds of years, so, too, does the idea of the self-regulating machine. In 250 BC, a Greek mathematician living in Egypt, Ktesibios of Alexandria, built the world's first self-controlling device. Ktesibios's creation was a water clock, featuring a regulator that meant that it maintained a constant flow rate. The clock worked by way of a float sitting in a jar of water. As water dripped out of a hole in the bottom of the jar, the float fell along with the water level. With each unit of passing time, a doll-like figure on the top of the float operated a gear mechanism. Depending on which of Ktesibios's clocks you examine, this resulted in either the dropping of a pebble or the sounding of a horn.

Ktesibios's water clock was significant because it forever changed our understanding of what a man-made object could do. Before Ktesibios's clock, only a living thing was thought to be capable of modifying its behavior according to changes in the envi-

---

*Coffee, as it turns out, is a good starting point for a discussion about smart devices. Apple's cofounder Steve Wozniak once said that he could never foresee a robot with enough general intelligence to walk into a strange house and make a cup of coffee. Exploring this hypothesis, some researchers now suggest the "coffee test" as a potential measure for AGI, Artificial General Intelligence. I will discuss AGI later on in this book.

ronment. After Ktesibios's clock, self-regulating feedback control systems became a part of our technology.

In the twentieth century, an influential AI pioneer named Norbert Wiener worked to formulate mathematical theories around feedback systems. Wiener proposed the idea that intelligent behavior comes about as the result of receiving and processing information: a concept which came to be known as cybernetics. During World War II, Wiener's theories regarding feedback systems were refined when he and a colleague named Julian Bigelow worked on a project designed to improve the accuracy of anti-aircraft guns. Wiener and Bigelow solved the problem of how to fire more accurately at a moving plane, which posed a challenge since it meant that the person doing the firing had to anticipate the future position of a target. Their solution was a method for automatically correcting a gunner's aim by predicting where the target was going to fly and adjusting the gun's sights accordingly.

Wiener's ideas about sensing and feedback as a way of optimizing performance weren't purely limited to warfare. More than anyone who came before him, Wiener imagined feedback as something that could be applied as a fundamental universal principle. Feedback, he believed, could be applied to machines, organizations, cities and even the human mind in equal measure. He recorded many of these ideas in a 1950 book entitled *The Human Use of Human Beings*, published six years before the official formation of AI. A surprise bestseller, *The Human Use of Human Beings* describes the various ways in which smart automation could improve society. Diverging from the idea of building thinking machines to replace human beings, Wiener used his book to discuss the various ways in which humans and machines can cooperate. In the introduction, he writes:

It is the thesis of this book that society can only be understood through a study of the messages and the communication facilities which belong to it; and that in the future development of these messages and communication facilities, messages between man and machines, between machines and man, and between machine and machine, are destined to play an ever increasing part.

Cybernetics never became the mainstream object of research funding that AI did. However, the idea that mathematical feedback systems could be used to forecast the future is the basis for almost all of today's smart devices. For instance, a regular "dumb" thermostat receives the temperature from a sensor, and based on how hot or cool it is, turns on your furnace or air conditioner. A "smart" thermostat, on the other hand, has the ability to incorporate other data sources, such as the day's weather forecast or knowledge about the historical patterns of room usage by people in your house. It could even select an average temperature based on the collective body sensor readings of a large number of people in a room. Instead of being simply reactive, smart devices become predictive.

This requires interactions between different devices. Smart devices might be smart compared to their pre-connected counterparts, but they are still nowhere close to having what we might realistically term "intelligence." But new possibilities are opened up when devices are able to share data and goals with one another. This is what experts describe as "ambient intelligence," referring to the way in which multiple devices can act in unison to carry out tasks by using the intelligence embedded within a network. Like termites working en masse to build a termite mound, the whole is greater than the sum of its parts.

The Dance of the Tortoises

This interest in emergent behavior between machine and environment (or, better yet, multiple machines and their environment) was another insight that came out of the cybernetics movement. It triggered some important early work in robotics, such as that carried out by William Grey Walter, an American-born neuroscientist living in England. In 1949, Walter built the world's first pair of three-wheeled robots, which he referred to as "tortoises." Unlike the computer scientists just beginning to work with digital computers, Walter relied on analog electronics to simulate brains for his robots. His aim was to prove that rich connections between a small number of brain cells could give rise to complex behavior. He was fascinated by the notion that a machine might define a goal and then seek to complete it by learning from the consequences of its own actions.

Named Elmer and Elsie, Walter's tortoises each boasted a light sensor, marker light, touch sensor, propulsion motor, steering motor and protective shell. Although they were unreliable in their workings, the pair were capable of autonomously exploring their environment. In his book *The Living Brain*, Walter recalled how an elderly woman, feeling pursued by the free-roaming machines, fled upstairs and locked herself in her bedroom. The tortoises were later modified with the help of Walter's technician, W. J. "Bunny" Warren, at the Burden Neurological Institute in Bristol, where Walter worked. Three of the resulting "Machina Speculatrix" were shown at the 1951 Festival of Britain, featuring several notable improvements over the original Elmer and Elsie. This included the ability to steer toward a light source when the robots were running low on battery power. Largely forgotten today, Walter's tortoises nonetheless serve as early examples of autonomous robots able to learn through trial and error from their own actions and mistakes.

It is impossible not to look at a device like the Roomba robot vacuum cleaner, created by the iRobot Corporation, and see the legacy of William Grey Walter's tortoises. The Roomba is a small, puck-shaped, computerized vacuum cleaner that automatically guides itself around your home. It follows a series of pre-programmed cleaning strategies, although it can also respond to stimuli through feedback-based "intelligence." At first, the Roomba follows two simple instructions: "wall following" and "random bounce." In the first mode, it follows the walls in a room using a side-mounted brush to clean into corners. In the second, it cleans until it collides with an obstacle, at which point it changes course and heads off in a new direction. To help it move efficiently, the Roomba contains a number of smart sensors. Two of these are infrared beams, which help it detect walls and what it calls "cliffs," referring to stairs and other drops. A touch-sensitive bumper stops the Roomba from moving forward when it hits another obstacle. On its underside it also has what is known as a "piezoelectric sensor," allowing it to detect bits of dirt. If it discovers too many pieces of dirt in one place, the Roomba retraces its steps and cleans a second time, this time slower and more thoroughly. Observing just these simple steps, the device exhibits a sort of emergent behavior that can appear almost lifelike.

In some sense, the word "emergent" suggests that the behavior is not understandable. This is not true. Based purely on the simple rules laid out above, we can understand why the Roomba behaves in the way that it does. However, as with Walter's tortoises, the combination of behavioral agent and environment can bring about some unexpected responses as the Roomba seeks to fulfill its task.

One Roomba is all well and good. But as Walter discovered with his tortoises, it is when you have more than one agent interacting that things get really interesting. Among Walter's most intriguing observations was his discovery of the way his tortoises "danced" when they were around each other. This dance consisted of a kind of seemingly ritualized bumping and backing up on the part of each robot. It was the result of the marker light he had attached to each tortoise, which was activated when the turning motor came on, but went out when the turning stopped. As each tortoise oriented itself to the other's light, they attracted each other like two creatures of the same kind meeting for the first time. A similar act occurred when the tortoises passed a mirror in which they were reflected. Walter claimed that, were it an animal behaving in this way, it "might be accepted as evidence of some degree of self-awareness."

Even Roomba enthusiasts are unlikely to go so far as to suggest that two robot vacuum cleaners interacting with one another have "self-awareness," but Walter wasn't wrong in suggesting that multiagent systems is where smart devices become more interesting. For instance, what would happen if the doors in your house were able to open and close automatically to allow your Roomba to clean more than just one room at a time? This might be desirable in some situations, such as if you had a pet you didn't want to let into a certain room or if you were heating a particular room. Similarly, what if your Roombas had access to a sensor on your front door, or your car, and knew to start up the second you left for work so that the cleaning was finished when you got home? Perhaps unsurprisingly, this is the direction some of the biggest companies building smart devices are now headed in.

Your Biometric Biographer

Until June 2015, I'd never spent too much time wondering about which city's residents get the least amount of sleep, or whether commuters who travel less than five miles to work exercise more than those who travel farther. The answers, for those who are interested, are citizens of Tokyo, Japan (five hours and forty-four minutes), and "yes" (by 422 steps per day).

I know this because Bandar Antabi told me. Bandar is surely a contender for the world's best ever pub quiz contestant. Ask him and he can tell you that the best place to live if you fancy an early night is Brisbane, Australia, where people tend to fall asleep at around 10:57 p.m., while the best place for night owls is Moscow, Russia, where shut-eye usually occurs at 12:46 a.m. Women eat 3 percent less garlic than usual on Valentine's Day, he says, but men make up for it by consuming 37 percent extra. Folks in Stockholm, Sweden, are the world's most active walkers on a per-day average, while those in São Paulo, Brazil, are the world's least active. And so on, as if you had set Dustin Hoffman's character in *Rain Man* loose on Wikipedia for a few hours.

Bandar's a smart guy, but the truth is that he's not actually unusually good at trivia. He can reel off all of this information because Jawbone, the company for which he works as Head of Special Projects, has spent the past several years tirelessly collecting it.

Jawbone started out making noise-canceling technology for the US military in 1999, before expanding into consumer Bluetooth headsets, speakers and, eventually, wearable lifestyle track-ers. It is this last category of sensor-filled smart devices—such as the UP3, a thin, watch-style band that obsessively catalogs every-thing from your sleep patterns and respiration to your heart rate and "galvanic skin response"—for which Jawbone is best known today. The raw data generated by Jawbone's army of users is what has given Bandar his plethora of factoids. At present, that data in-cludes some 3 trillion steps, more than 250 million nights of sleep, and close to 2 million meals. As time goes on, it will expand to include potentially dozens of other metrics, such as the amount of caffeine its users drink on a daily basis. In short, Jawbone wants to be your biometric biographer.

"Our mission is to build this personalized data set that has your identity, your profile, your biometric information, your age, your height, your weight, your gender, your food preferences, your mood," Bandar tells me, as I sit opposite him, sipping a Starbucks coffee on the thirteenth floor of a Notting Hill office block, which serves as Jawbone's UK headquarters. "We want to know your as-sociated activities as well. When are you sitting? When are you active and burning calories? What's the quality of your sleep? By plotting this over time, we can tell you an enormous amount. We're building a contextualized data set around your well-being."

Jawbone has brokered deals with plenty of other tech hardware players, but these wouldn't be worth the paper they're written on if it was simply about sharing data. Does your thermostat really need to know what you ate for supper last night? What good can it pos-sibly do if your television is aware that you like to go jogging four mornings a week? Actually it means a whole lot, Bandar says. "Data is good," he tells me. "But understanding data is what we're all about."

It is this "understanding" which means your data can be parsed by the proper AI algorithms in a way that makes contextual sense. "We can use this technology to push data to the correct devices in a way that is useful," he continues. "For example, you can pair your Jawbone device with your smart thermostat so that when you fall asleep the temperature in your bedroom is adjusted to one conducive to your optimal sleep patterns. When you wake up, the temperature changes again."

These kinds of data transactions are known as event-driven programming, or "If-This-Then-That" rules. These are simple rules for stringing together chains of services based on simple recipes. "If-This-Then-That" pioneer Linden Tibbets has described such rules as "digital duct tape," since they allow the creators or users of smart technologies to connect otherwise entirely separate concepts together. There are countless other examples of how such interactions may, or currently do, work in the world of smart devices. For instance, if your car knows that you didn't sleep well last night, it could draw on data from your smart thermostat, revealing that you respond positively to the cold. It might then crank up the air-conditioning to ensure that you are as alert as possible. It may also be aware, from analyzing your fitness-tracking wearables, that you perform best when you listen to a certain genre of music. As a result, it could automatically play Metallica to get you revved up for the day. It may even ascertain that you are still drunk after going out with your buddies the night before. To establish this, it uses smart sensors embedded in the gear stick, analyzing the alcohol content in the sweat of your palm. In that case, it could shut off the car completely and recommend that you call an Uber cab.

As another illustration, your smart TV might access your sleep records and suggest customized viewing schedules based on the time of day. Instead of sitting down at 9:00 p.m. to watch a show like *Game of Thrones*, which is guaranteed to keep your brain buzzing for hours, why not cue up *Modern Family* instead? Or perhaps you were watching TV and saw a recipe you liked, which could easily be sent to your smart fridge. Because the fridge monitors its contents, it knows if it has the necessary ingredients for you to make the dish. If it doesn't, it can add them to your next grocery list for home delivery. As more and more of these devices are networked together, able to draw on one another's data and linked by event-driven programming, a long-term techie dream will start to be realized.

I'm referring, of course, to the existence of the fully fledged smart home.

The House of the Future

When I was growing up, it seemed that every TV show sooner or later featured an episode based around the idea of the "house of the future." One of my favorite such episodes came from the BBC sitcom *Some Mothers Do 'Ave 'Em*, centered around the mishaps of wimpish, accident-prone Frank Spencer. In the episode "George's House," Frank and his long-suffering wife, Betty, visit Betty's brother George, a high-tech designer who lives in a home filled with the latest smart gadgetry. Things inevitably go wrong. While attempting to use a sensor-filled toilet, Frank accidentally breaks the flushing mechanism. When he tries to fix it, a toilet brush, ball cock, and pair of slippers quickly become wedged in the toilet. By the end of the episode, the home resembles a haunted house: doors and windows slamming open and closed apparently of their own accord. The head of a building firm who has misguidedly chosen the day of Frank's visit to be given a tech demo of the home flees in abject terror. "Get me out of this crazy house," he screams, shortly before the gates in front of the property are blown off their hinges in one final laugh.

In reality, our houses have been steadily getting smarter for years. For the 1933–4 Chicago World's Fair, the modernist architect George Fred Keck designed a "House of Tomorrow" exhibit. It included a built-in dishwasher, electric garage door opener, central air-conditioning and electric lights, complete with dimmers. All of these were considered almost implausibly futuristic for the time. Now, we have already moved on to the next evolutionary step. We don't necessarily think of it this way, but even a mid-level security system that calls the police if a motion sensor is triggered is smart technology. Where mass-produced electrical goods were Keck's vision for the future, today similar scenarios are based around smart devices.

Picture, if you will, yourself in 2020. You arrive back from work at six o'clock, and as you roll up the drive in your car, the garage door ahead of you opens silently, beckoning you into your ultra-connected smart home. You enter your house, and immediately the lights turn on, configuring themselves to your favored low-light setting, while the smart thermostat sets the temperature at a comfortable 73 degrees. If there are multiple people who live in the house, your home knows that you are "you" through some kind of biometric authentication technique like fingerprint sensors or facial recognition. In the background, your Wi-Fi speakers immediately spring to life, having selected some heart rate–slowing soft rock music to play. Your evening meal is almost ready, since your sous-vide immersion cooker knew when you left the office based on your geo-location, so there's just enough time to take off your jacket and pour yourself a beer. In the corner of your sitting room, the 55-inch smart television set says, "Welcome home," in its familiar voice, and suggests that you might enjoy watching the highlights from last night's football game, which it knows you haven't seen yet.

In addition to devices being able to talk with one another, one of the big differences with the smart connected home will be the use of Artificial Intelligence to form goals that our gadgets can work toward to make our life easier, more comfortable, or more productive.

"The idea of the Internet of Things, all these devices that are thinking a bit, could go one of two ways," says Richard Sutton, an expert in a field of AI called "reinforcement learning," which deals with AI capable of forming and pursuing goals. "You may have isolated agents which behave with our own localized goal. For example, your thermostat's 'goal' might be to be efficient and not use too much fuel. Your refrigerator's 'goal' is to make sure that it's fully stocked to serve you food whenever you want it. The result could be that your smart devices fight among each other to work out which device's goal takes top priority. The alternative would be to have all of them interconnected to become one decision-maker."

It's not difficult to imagine similar examples—such as having a house that prompts you to get fitter or cut down on fuel usage. Not everyone will like this, of course. "There's a certain clarity to the first option," Sutton continues. "It means that you know your furnace turns on because it's cold, not because it wants to keep you indoors so your smart TV can play you the latest show. It's got a clear goal and you both know what it's working toward. Its credit assignment for itself is very straightforward."

But so long as they are presented in a way that is transparent to the user, there is also a case for larger macro goals that allow your devices to work together toward longer-term, more complex ambitions.

Home Is Where the Sensors Are

Diane Cook is a professor at Washington State University's School of Electrical Engineering and Computer Science. For the past several years, she has been investigating ways that smart homes can improve the life of elderly people. A few years ago, Cook visited the Texas State Fair, where she saw a "home of the future" exhibit featuring a range of smart devices. She came away unimpressed. "I looked at it and thought, 'That's not a smart home, that's a connected home,'" she says. "There was a refrigerator that let you scan in barcodes, which would then maintain a grocery list for you and send it off to your local food store, who would deliver it. It had a lot of those kind of neat devices—but the 'smarts' were still the burden of the person who lived there. There was no reasoning involved. It was just information."

Using her knowledge of machine learning, Cook wanted more than a house that would collect data. "I view a smart home as one which not only senses what is going on in the environment, but can also act on the environment through automation," she says. "It reasons about the information that it's sensing, and uses that information to intelligently select an action for automation." Cook began work on a project designed to use this smart reasoning process to identify early indicators of cognitive and physical difficulties in elderly people. Smart sensors can be used to tell us a whole lot about what a person is doing in their home. Devices such as infrared motion detectors, magnetic door and window alerts and sensors that can track the status of water taps and stovetops can reveal whether a person is eating, sleeping, cooking, watching TV or going for a walk. By monitoring the activities that are attempted and extracting "statistical activity features" about how they're performed, Cook's algorithm can predict how well a person is coping. For instance, a person having memory difficulties may take longer to perform certain tasks. They might also exhibit telltale signs like wandering around more, opening and shutting cupboards while trying to remember what is next in a particular sequence of events or using an incorrect tool for cooking. In isolation, these behaviors don't have to mean anything, but taken in aggregate they paint a revealing picture.

Cook and her team began testing this smart home technology on campus at Washington State University. They then moved on to a local care home, Horizon House in Seattle. A total of eighteen seniors, aged seventy-three and above, volunteered to be part of the study. They had their apartments fitted out with sensors, which took the form of small white boxes, measuring around one inch by two inches. Even without cameras, the sensors were able to discern between two individuals living together, or a human and a pet— the latter of which Cook describes as a "nightmare scenario to have to deal with from a smart home perspective." Diane Cook then compared the data gathered from the sensors to the regular check-ups on residents administered by (human) care workers. "It was surprisingly successful," she says. "We found a very high correlation between how people perform activities and what health diagnosis category they were in. As a result, we were able to successfully predict from the machine learning tools what their diagnosis would be, purely based on how they perform a few activities."

According to Diane Cook, while not intended as a substitute to social visits from friends and family, this technology will enable an elderly person to live independently for longer—which might mean staying on in the house they've spent much of their life in. "That would be possible even if there are not caregivers or family right there on the doorstep. Caregivers or medical personnel could be alerted if there is a significant change which could equal a transition to a different health status." The onus isn't entirely on carers, either. The smart home can alert individuals of "senior moments," such as leaving the refrigerator open or the stove on.

Having received $3 million funding for the project, Cook's next goal is to expand the study. "Today we have a smart home in a kit," she says. "We can put all the sensors, networking, software and computer into a small tub and ship them to sites around the world." Before long, it should be possible to carry out international studies—with the machine learning algorithms getting ever smarter as data flows in from around the globe.

A number of companies are also working in this area. For example, Healthsense makes the eNeighbor monitoring system, a wearable device augmented by a range of smart sensors for the home. A bit like Diane Cook's Horizon House project, eNeighbor can be used to detect falls or alert a caregiver if a patient has forgotten to take their medication. Similarly, the BeClose smart sensor system notes lengthy absences or missed meals and sends a text message, e-mail or phone call to designated family members.

Having smart home information is all well and good, but the next wave of smart devices will additionally allow for the tracking and diagnosis of diseases. Take the medical device maker Alive-Cor, which manufactures a smartphone case that doubles as a portable EKG heart monitor, able to predict whether a user is about to suffer a stroke. The case works by measuring the heart's electrical patterns through the fingertips of the person holding it. An algorithm then analyzes the regularity of their heartbeat and suggests if the person should see a doctor.

As our environment gets ever smarter, we will enter an age of continuous, real-time risk assessments. For the first time in history it will be possible to draw constant correlations, and possibly causations, between a large number of genomic, physiological, biological and environmental factors on an individual basis. Wearable devices will tirelessly monitor our heart rate, blood oxygen levels, physical activity, breathing patterns, facial expression, lung function, voice inflection, brain waves, posture, sleep quality and more, in addition to external measurements like air quality and noise level. Using insights from Artificial Intelligence, these data points won't merely be turned into generalized advice about your life as a whole, but rather actionable insights capable of improving health on a moment-to-moment basis. Carrying out both prediction and diagnosis, we will learn exactly what conditions are necessary for a particular illness or episode to occur, and can make proactive preventative steps to ensure that these do not happen. An asthma sufferer could have the specific triggers responsible for an attack—perhaps cold, exercise, pollen, or some other allergen—analyzed by their smart devices. When these conditions risk repeating again they may be warned to take their medication early, or else avoid a specific location. In another example, an unknowing sufferer of the neurodegenerative disorder Parkinson's might be alerted of the disease's onset long before a doctor could diagnose it, by way of subtle vocal tremors and reduced speech volume unnoticeable to the human ear. Although there is currently no cure for Parkinson's, an early diagnosis may help improve quality of life.

While the data in all of these cases would be available to the user, it would not be necessary for them to see it unless there was cause for concern. For instance, a health-tracking technology's default position might be a high-level instruction stating, "Monitor my vital signs and, so long as they're okay, don't transmit anything." If a potentially significant change is noted, the system will alert the user or, in other instances, their doctor.

This kind of technology is new in the medical realm, but common in other parts of our lives thanks to machine learning. For example, algorithms are currently used by banks for fraud detection. Although we have the ability to look at every transaction that takes place within our bank accounts, we are only alerted when the bank notices behavior that deviates from our regular patterns. If I typically make regular payments of less than $194, but then make a one-off online payment of $1,500, it is likely to be flagged as suspicious. Machine learning is also used by many e-mail systems to sort "spam" or junk e-mails from ones that we want to read. Spam filters work by applying a score to each incoming e-mail, based on a series of built-in rules. These scores are honed over time as the spam filter observes how we respond to the different messages we receive. We are only shown e-mails that conform to the score our spam filter deems worth reading.

The City That Never Sleeps

Similar smart technologies promise to transform our cities, too. The growth of cities and the spread of information technology have always been intimately linked. In 1910, the historian Herbert Casson wrote, "No invention has been more timely than the telephone. It arrived at the exact period when it was needed for the organization of great cities and the unification of neighbors." With their interconnected web of technologies, all working together to output wealth and productivity, cities have regularly been portrayed in media as living, breathing entities. This is most evident in a film like Fritz Lang's 1927 sci-fi masterpiece, *Metropolis*, where the titular metropolis is depicted as an enormous living organism.

The dream of Artificial Intelligence appeared to take us one step closer to this possibility. In 1964, the same year as the New York World's Fair, the British architect Ron Herron came up with his concept for a "Walking City." Described in the avant-garde architecture journal *Archigram*, Herron argued for the construction of enormous, artificially intelligent mobile robotic platforms capable of roaming the Earth like giant skyscraper-carrying spiders. These walking cities would exist in borderless worlds in which they were free to go wherever they needed to acquire the necessary resources or manufacturing abilities. Herron's cities would even, he explained, have the ability to connect with one another to create even larger "walking metropolises." Not only would such cities be self-sufficient but, thanks to breakthroughs in AI, literally autonomous.

Given the state of robotics research at the time, it is probably for the best that Ron Herron's ideas were never taken seriously. SRI's SHAKEY robot, as described in chapter one, was incapable of so much as maneuvering a hallway without running into problems, so it's no small mercy that we were spared the sight of an autonomous New York toppling over after encountering slightly uneven ground. It is unlikely that cities will change quite to the extent that Herron proposed, but they will certainly continue to get smarter. For instance, smart offices will feature sensor-filled wastepaper baskets that alert maintenance staff when they need to be emptied. Employees won't have to be aware of health and safety regulations, since the office will continuously monitor its own temperature and compare these to the levels stipulated by law. If a level is exceeded, an alarm may sound while computers automatically turn off. In shops, bars, theme parks and museums, Bluetooth beacons will transmit user-relevant information to your smartphone or wearable device depending on your location and personal preferences.

On the street, by far the biggest visible change likely to happen in the next several decades will be the mass arrival of self-driving cars. Following on from the work of Dean Pomerleau, as described in the last chapter, both Google and Apple have invested in this field and look set to play a key role in bringing autonomous vehicles to the mainstream. Self-driving cars won't only affect us on an individual level, but also collectively by helping to reduce traffic congestion in cities. The data that they gather will be vital to town planners as cities continue to expand. We are already starting to see how this may work. In early 2015, the Google-owned traffic app Waze teamed up with the city of Boston to reduce local traffic. Boston agreed to give Waze advance notice about planned road closures, while Waze agreed to share the app's valuable stream of data with the city's traffic management center. Short-term, the collaboration made Waze more efficient at helping users to reach their destinations quickly. Long-term, the idea is that Waze data will help the city fine-tune its traffic-light timings and work out how to cut down on congestion.

Get Ready for the Internet to Disappear

In January 2015, Google's executive chairman Eric Schmidt caused a stir while speaking at the World Economic Forum in Davos, Switzerland. Asked about his predictions for the future of the web, Schmidt said, "I will answer very simply that the Internet will disappear."

There was, of course, nothing simple about this answer. Upon first listen, it was a bit like Apple CEO Tim Cook telling people that they should put down the smartphones and have a face-to-face conversation with friends, or a movie studio boss saying that cinema is stuck in a rut and people ought to spend their time reading books or going for walks. In reality, Schmidt was saying nothing of the kind. Instead, he was making an observation about what has happened to technology in recent years as it has become both smaller and more pervasive. He wasn't the first person to make such a suggestion. In 1991, Mark Weiser, chief technologist at the legendary Silicon Valley research lab Xerox PARC, wrote an article about what he termed "ubiquitous computing." It opened with the lines: "The most profound technologies are those that disappear. They weave themselves into the fabric of everyday life until they are indistinguishable from it."

It is hard to argue that this has not been the case. ENIAC, the groundbreaking digital computer described in chapter one of this book, weighed 60 pounds and took up an entire room. A regular clamshell cellphone made early this century (not even a smartphone) is approximately 120,000 times lighter than ENIAC, uses 400,000 times less power, but still manages to be 1,3000 times more powerful. This transition will continue as wearable devices take over much of the functionality currently found in our smart-phones. In the same way that laser discs look archaic to us in an age of Blu-rays, so first-generation wearables will appear laughable just a few years down the line. Google has already developed smart contact lenses capable of measuring the glucose levels in a wearer's tears and then transmitting this information wirelessly to a con-nected smartphone. In the process, the search giant hopes to re-move the need for diabetics to perform regular, often painful blood tests. On an even smaller scale, another company, Scripps Health, is working to develop a nanosensor that users can inject into their bloodstream. Once there, it will nestle into the body's capillary beds that supply blood to the body's various organs, pick-ing up relevant readings to transmit back to a master device for analysis. Users won't even have to worry about how the sensor is powered, since it has the ability to act as a mini hydroelectric dam by using the force of the blood that passes by.

These technologies are virtually invisible to the human eye, but more important, will also be invisible to the wearer. At present, smart devices still require us to do a number of things manually, meaning that we are not quite yet in the realm of seamless smart interactions. Jawbone, for instance, has a comparatively small data set of meals to analyze compared to nights of sleep. This is because people currently have to log meals themselves, selecting each item from a long list one at a time. If it were possible to photograph a meal and then have the image recognized as, say, an omelette and chips and logged accordingly, it is far more likely that people would do it. This is an area tech companies are exploring. In 2010, the tech giant Qualcomm patented a technology allowing users to pair devices—such as a smartphone and smartwatch—by simply snapping a photo of them. Once image recognition tools identify the new device, the two machines interface to automatically con-figure the pairing process. This is much simpler than the complex manual pairing process, which is why it is now being used by sev-eral smartwatch makers. It can only be so long before similar tech-nology can be linked to a food database.

"This is certainly something tech companies are working on," says Michael Grothaus, the SITU smart scale entrepreneur from the start of this chapter. "Right now, we're seeing some really inter-esting work done with devices called spectrometers, which use light to measure the composition of materials. The problem is that they still can't accurately read both the makeup and mass of an object. One day they will be small and cheap enough that we can measure everything with ease, but until then the best way to mea-sure the calories in food is to manually log it."

The dream of technologists like Grothaus is to make smart de-vices invisible not only in size but in their use. Just as we do not have to consciously think about our heart rate, body temperature or breathing in order for the central nervous system to regulate and control it, so too smart devices will increasingly gather and communicate information without individuals having to be re-sponsible for the monitoring process.

The Problem with Smart Devices

This brings with it a number of ethical dilemmas. The problem with a device that is "invisible" is that we can miss some of the finer points of how it operates. Particularly if it is a device with a pre-smart analogue equivalent, we may assume that it works in exactly the same way. As an example, we've already discussed how smart devices are capable of working toward both micro and macro goals. However, these don't necessarily have to be goals that only benefit you, the user—even if it is you who owns the smart device in question. Insurance companies, for instance, have shown themselves to be keen to use connected smart devices as a means by which to optimize security premiums. Put simply, if you're healthier and safer, your premiums go down. Currently, insurance rates are calculated on an annual basis, making room for any changes in your circumstances. Using smart devices, rates can be adjusted continuously, with premiums going up and down like stock prices according to whatever your latest readings show.

Big businesses are embracing the use of wearables as a way of tracking staff. The oil company BP has given 14,000 of its employees free Fitbit Zip activity trackers on condition that they allow the company to look at the number of steps they are taking. Called the "Million Step Challenge," BP employees who walk more than 1 million steps are rewarded with lower insurance premiums. According to Fitbit, employees who use Fitbit devices in similar corporate wellness programs take 60–80 percent more steps than the average person. In some ways, this works out well for all involved. Companies pay lower insurance costs for their employees, employees are healthier and national health care costs are reduced. The research firm CDW Healthcare has reported that wearable technology could reduce hospital costs by as much as 16 percent over the course of a five-year period.

But with this comes the threat of a more Orwellian society. In particular, it is reminiscent of "Taylorism," an early twentieth-century movement pioneered by the engineer Frederick Taylor. In Taylor's influential 1911 book *The Principles of Scientific Manage-*

ment, he laid out his beliefs that the purpose of human work and thought should be increased efficiency. Taylor carried out studies designed to teach employers about how to measure the previously unmeasurable to increase their profits. For instance, in his "science of shoveling" experiment, Taylor determined that the optimal amount of weight a worker should lift with a shovel was precisely twenty-one pounds. By doing this, efficient shoveling speed could be maintained for longer. This is exactly the kind of thing that could now be easily measured by a smart device and fed back to your boss. Amazon today uses similar technology in its factories, where "fulfillment associates" (a.k.a. product pickers) are issued handheld computers that record how fast they complete individual orders. Taylor's ideas of scientific management didn't only favor employers. He strongly believed that the ability to measure work should also go hand in hand with incentivized remuneration, so that a low-performing employee with poor productivity does not make as much as a higher-performing one. All this makes perfect sense in theory, but critics point to the fact that scientific management also reduces autonomy, penalizes the old, weak, or disabled and, in an ironic twist given the topic of Artificial Intelligence, treats men like machines.

In other cases, we might be lucky if certain aspects of our devices are designed to benefit us at all. In 2014, two writers for *Forbes* magazine revealed that smart device maker Nest has deals with electricity companies to provide data gathered from its users about their habits. While the data is anonymized and only reported in aggregate, it is still used by power companies to manipulate the smart devices within our own homes. To ease the load on its grid, power companies can ask Nest to turn down users' air conditioners on a hot day. Nest splits the cost savings with the utility company—with users getting nothing. Over time, Nest's revenue from deals with utility companies will dwarf the amount it makes from sales of its thermostats. The company's smart devices are still serving their master; it's just a different master from the one we may have expected.

Related challenges may be faced as user data is gathered by smart devices and used to shape cities. Instead of smart cities becoming increasingly cohesive, they could be made more divided, depending on how AI is employed. One deep learning project created at MIT's Computer Science and Artificial Intelligence Laboratory (CSAIL) found that it could predict the crime rate in an area simply by looking at an image. Trained on 4 million images from Google Street View in addition to aggregated crime data from organizations like San Francisco CrimeSpotting, the deep neural net focused less on what was present in a particular image and more on inferences. "What we're trying to do is show that studying images should be about more than just analyzing what is visible," Aditya Khosla, one of the project's creators, told me. "If the goal of Artificial Intelligence is to build machines that can mimic human intelligence, this level of abstraction is the obvious next step." Like many of the applications described in the last chapter, CSAIL's project was an impressive example of deep learning in action. But how it is used is open to human interpretation. For example, town planners could use the neural network to pinpoint parts of a city that desperately need investment, or where a hospital or school is needed but is not currently built (another use of the neural network). A car company, meanwhile, may use the same technology to trigger the doors automatically locking in your vehicle or the selection of an alternate travel route, thus reinforcing the sense that this area is a crime "ghetto."

With so much to think about, you wouldn't be blamed for wanting to turn some of this over to a digital proxy, or a smart assistant you can trust.

Fortunately, AI can help in that capacity, too.

4

How May I Serve You?

NEGOBOT IS, FOR all intents and purposes, a fourteen-year-old girl. Her speech is often bored, her interests made up of popular bands and clothing labels. Her writing is full of Internet slang like LOL ("Laugh Out Loud") and peppered with emojis: the little cartoon smiley faces which serve as emotional shorthand online. Sometimes she sounds surprisingly adult. Other times she is most definitely a child.

She is trying her hardest to work out whether or not you are a pedophile.

Negobot is the creation of a group of researchers at the University of Deusto, Spain. She (or, more accurately, it) is an intelligent agent designed to mimic the speech and behavior of a young teenager online. In a world in which young people spend more and more time communicating online, the goal of Negobot is to act as a digital undercover agent by going into Internet chatrooms and seeking out suspicious individuals.

"Negobot is designed to go after pedophiles who are very difficult to catch," says Carlos Laorden, a researcher at the University of Deusto's wonderfully titled Laboratory for Smartness, Semantics and Security. "These individuals will typically groom their victims through conversations lasting several months. It takes an enormous number of man hours to police chatrooms for that very reason. The idea for Negobot is therefore to be able to simulate a human conversation not just for a few minutes, but for a sustained period of time."

Carlos Laorden started out his career creating programs for filtering out spam e-mails, a classic machine learning problem. By trying to find and isolate malicious behavior online, usually based on the language involved, he came up with what is possibly the most advanced real-world version of that ambition.

Negobot is programmed to operate according to the rules of game theory. Game theory was a concept first suggested by the math pioneer John von Neumann, whose work I briefly described in chapter one. It is the study of strategic decision-making, in which there are multiple players all with their own motives. The payoff depends on the behavior of these different players. Not everyone can get what they want—and the aim is to predict how people will act and hopefully to turn this to your advantage.

In the case of Negobot, the game's goal is to work out whether or not the person Negobot is speaking with is a pedophile. While it is doing this, it also wants to extract the highest amount of possible evidence against them, despite appearing only to passively respond to questions. Conversations with Negobot start out neutral, before "leveling up" according to the responses it is given. The AI has seven different levels of behavior in all, which it runs through over the course of the scenario. Each level corresponds to the perceived "sliminess" of the human correspondent. To begin with,

Negobot talks about its favorite films, music, personal style and clothing—as well as more suggestive subjects like drugs, alcohol and family issues. Depending on the way the conversation is led by the human participant, it can then expand to discuss sex and other taboo topics, while appearing to give out more "personal" information.

The unwitting human player thinks he or she is cleverly manipulating the conversation by discovering more about the "fourteen-year-old girl" they are supposedly speaking with. All the while this is going on, Negobot is building a case file against them.

"I can see this being a very useful automatic tool for identifying potential suspects," says Carlos Laorden. "If we use tools like Negobot, we can dramatically reduce the workload on the human teams currently working to catch these criminals."

Beating the Turing Test

Entrapment laws mean that Negobot is not currently being used by police forces around the world, but that doesn't make the experiment any less interesting. If anything, it serves to highlight just how broad the possible applications of conversation AI can be. At its root, Negobot offers a unique twist on the famous AI experiment known as the Turing Test.

Based on a hypothesis by Alan Turing, whose work I discussed in chapter one, the Turing Test is designed to test a machine's ability to show intelligent behavior indistinguishable from that of a human. As it is regularly performed, the Turing Test involves taking a computer (A) and a human (B), and having them each communicate with a human interrogator (C), whose job it is to figure out which of *A* and *B* is the human and which is the computer. If *C* is unable to do this, Turing argued that the machine has "won" and we must consider it to be intelligent, since we are unable to differentiate it from our own human intelligence. In the future, tools such as Negobot show that our ability to discern between real people and bots may even have legal ramifications.

No one alive today has done more to promote the idea of the Turing Test than Hugh Loebner, a colorful, self-proclaimed egotist with dyed black hair, who started out his career selling folding disco dance floors and is now in his seventies. The decision to stage what is now one of the world's best-known AI competitions only occurred to Loebner when he was well into middle age. The success of what he named the Loebner Prize is all the more surprising given that he has no qualifications whatsoever in computer science. "In years to come, there may be richer prizes, and more prestigious contests, but gads, this will always be the oldest," Loebner has said, with typical chutzpah.

Loebner started the annual Loebner Prize in 1990. Each year it draws "chatterbot"—or "chatbot"—creators from all over the world, as their conversational AIs compete in text-based conversations designed to fool human judges. The best-performing chatterbot receives $4,000, which is paid out by Loebner himself. Still-unclaimed prizes of $25,000 and $100,000 are promised to the first chatterbot judges cannot distinguish from a real human, and the first chatterbot judges cannot distinguish from a real human in a test which involves deciphering and understanding text, visual and auditory input. When this last goal is finally met, Loebner says the competition will end.

We are not at that point yet, however.

Here, for example, is my conversation with Mitsuku, a previous winner of the competition, which claims to be an eighteen-year-old female from Leeds:

Me: Hi Mitsuku.

Mitsuku: Hello there.

Me: How are you?

Mitsuku: My logic and cognitive functions are normal.

Me: I'll take that as a positive.

Mitsuku: Take everything I say with a grain of salt.

Me: Not a grain of silicon?

Mitsuku: If not that, what?

Me: I'm not sure, but I'm getting a little confused.

Mitsuku: Are you willing to find out? Are you looking forward to getting that?

Me: I feel we're getting somewhat off track.

Mitsuku: What is that feeling like? Are you going? OK, good evening, afternoon err. morning madame. BYESPLIT X1234|||||||||||||||||||||||||||||||||||||||||||||||| Do you think I am off track too?

Me: This is all a bit too *Terminator* for me, to be honest.

Mitsuku: I appreciate honesty. For you? You don't say.

Me: Are you familiar with *The Terminator*?

Mitsuku: Robots take over the earth in this futuristic sci-fi thriller.

Me: That would be the one.

Mitsuku: I remember.

Me: A friend of yours, perhaps?

Mitsuku: How well do you know this friend?

Me: Not as well as I thought.

Mitsuku: I don't know. How well does you thought?

Me: I have to go.

Mitsuku: See you later. Nice chatting with you.
    BYESPLIT X1234||||||||||||| UNKNOWN||||
    Terminator|||| Terminator| Terminator||||||||||||||||||||||||.

Not the start of a lasting friendship, perhaps. Mitsuku does, in its favor, possess some nifty abilities. In addition to pre-programmed answers, Mitsuku can recall references you have made earlier in a conversation, and perform limited "reasoning" in the manner of an expert system. Ask her if she can eat a house and Mitsuku will look up the properties for "house" in her internal database. Finding that the value of "made_from" is set to "brick," Mitsuku will answer that no, a house is not edible. But this reasoning remains inadequate for answering tougher questions like the one posed by Luciano Floridi, Professor of Philosophy and Ethics of Information at the University of Oxford, when he took on the role of judge at the Loebners one year. "The four capitals of the UK are three, Manchester and Liverpool. What's wrong with this sentence?" Floridi wrote. Mitsuku had no good answer.*

Not everyone is enamored with the Loebner Prize. Marvin Minsky called the competition "obnoxious and stupid." Part of this is down to Hugh Loebner himself, who seems to have an ingrained desire to upset the AI old guard wherever possible. Years ago, he provoked Minsky so much that Minsky finally snapped and said he would put up $100 to whoever could stop Loebner from staging his infernal contest. Loebner argued that since the

---

*To be fair to Mitsuku, very few of us would have a good answer if this question were put to us. As another prominent AI researcher, Yorick Wilks, pointed out to me, all one could really answer this question with is, "Eh? The UK doesn't have multiple capitals." All that differs is the way that we phrase our bafflement.

only way the contest could be stopped was for someone to win its $100,000 grand prize, Minsky was essentially cosponsoring the Loebner Prize. He wasted no time issuing a press release to say exactly that. Minsky spent years fuming about it.

But the other reason some (although not all) serious AI experts dismiss the Loebner Prize is that it is, essentially, a trick of the light. It is reminiscent of a magician who is praised not for his ability to perform genuine magic, but rather for his use of sleight-of-hand and misdirection to create an impressive illusion. "Unfortunately, the chatbots of today can only resort to trickery to hopefully fool a human into thinking they are sentient," one recent entrant in the Loebner Prize told me. "And it is highly unlikely without a yet-undiscovered novel approach to simulating an AI that any chatbot technology employed today could ever fool an experienced chatbot creator into believing they possess [artificial] general intelligence."

Turing wasn't particularly concerned with the metaphysical question of whether a machine can actually think. In his famous 1950 essay, "Computing Machinery and Intelligence," he described it as "too meaningless to deserve discussion." Instead he was interested in getting machines to perform activities that would be considered intelligent *if they were carried out by a human*. It is this idea that the MIT psychoanalyst and computer researcher Sherry Turkle talks about when she says that we should take computers at "interface value." But even with this proviso, however, chatterbots are not yet at the level where we could consistently mistake them for humans—as my conversation with Mitsuku proved.

The gulf between chatbot "intelligence" and human intelligence was highlighted once again in March 2016. That was when Microsoft introduced Tay, an AI that—like Mitsuku—was designed to speak and act like a teenaged girl. Tay exhibited age-appropriate behavior, such as employing millennial slang and chatting about pop stars Taylor Swift and Miley Cyrus. Users could interact online with "her" by sending a message to @tayandyou on Twitter. Microsoft's idea was that, as an advanced chatbot, Tay would have the ability to learn from interactions with real people in order to become smarter—or at least better at faking it. "The more you talk[,] the smarter Tay gets," Microsoft noted. The plan backfired. Online trolls immediately began bombarding Microsoft's AI with controversial messages in an effort to corrupt its blank slate of opinions. Within twenty-four hours of going live, Tay started tweeting pro-Nazi messages denying that the Holocaust had taken place. By the time Tay started advocating genocide and messaged one user that "HITLER DID NOTHING WRONG!" Microsoft stepped in to pull the plug. "Tay is now offline and we'll look to bring Tay back only when we are confident we can better anticipate malicious intent that conflicts with our principles and values," wrote Microsoft's head of research in a formal apology to everyone hurt by the AI's "offensive and hurtful tweets."

But despite these deeply embarrassing mishaps, it doesn't mean that chatbots can't be useful.

The Rise of Virtual Assistants

A few months earlier, in January 2016, Facebook CEO Mark Zuckerberg announced his latest New Year's resolution. As the cofounder of the world's biggest social network and with a personal net worth estimated at $46 billion, Zuckerberg had already achieved more than most of us could hope to in multiple lifetimes. However that hadn't stopped the youthful innovator from setting himself one New Year's resolution each year in order to, as he puts it, "learn new things and grow outside my work at Facebook." In previous years, Zuckerberg had sought to read two books every month, to learn Mandarin, and to meet a new person every day. In 2016, it was something different again.

"My personal challenge for 2016 is to build a simple AI to run my home and help me with my work," he wrote in a post that appeared, naturally, on Facebook. "You can think of it kind of like J.A.R.V.I.S. in *Iron Man*," he added, offering us a handy pop culture reference.

It was a bold pronouncement and, at the time of writing, we have yet to see the end result. Zuckerberg's "personal challenge" looked to be the first time he had created a New Year's resolution that would be unavailable to the rest of us. After all, by likening his plan to Iron Man's AI butler J.A.R.V.I.S., it was a real-life billionaire referencing the creation of fictitious billionaire Tony Stark. It was a bit like Elon Musk announcing that he planned to use his fortune to build a fully working version of *Star Trek*'s USS *Enterprise*.

In fact, over the past five years, functional, AI-driven chatterbots have increasingly become part of our daily lives. Most famous of these is probably Siri, the Apple-owned AI assistant that first shipped with the iPhone 4s in late 2011. Using Siri, iPhone owners can ask natural language questions like "What is the weather today?" or "Find me a great Greek restaurant in Palo Alto" and receive accurate spoken answers.

Siri's abilities extend way beyond those of the chatterbots I witnessed at the Loebner Prize, although it is also programmed with enough nonproductive chatter that it is fun to speak with. Ask Siri for the meaning of life, for instance, and she will answer "42" in a geeky reference to Douglas Adams' *The Hitchhiker's Guide to the Galaxy*. Proclaim that "I am your father" in a reference to *Star Wars* and it will respond, "Together we can rule the galaxy as father and intelligent assistant!" When Steve Jobs first got his hands on a finished iPhone 4s his first Siri question was reportedly, "Are you a man or a woman?" (Siri gained Jobs' stamp of approval by answering, "I have not been assigned a gender, sir.")

What makes Siri different to—and far more useful than—chatterbots like Mitsuku is its ability to answer useful real-world questions. For example, one of Siri's methods of answering knowledge questions is Wolfram Alpha, a tool developed by the British mathematician and scientist Dr. Stephen Wolfram. Wolfram Alpha comprises around 15 million lines of Mathematica code. Unlike a regular search engine, which provides users with a list of documents or webpages it thinks contains the answer to a query, Wolfram Alpha answers questions through computation. Quiz it on the number of primes in 1 million (78,498) or the country with the highest GDP (Monaco), and it will answer the question by actually working it out.

In other cases, Siri's reasoning allows it to extract the relevant concepts from our sentences and connect these with web-based services and data, applying its ever-growing knowledge about you to a series of rules, concepts and contexts. The result is a way of turning requests into actions. "I want to eat in the same restaurant I ate in last week," is a straightforward enough sentence, but to make it into something useful, an AI assistant such as Siri must not only use natural language processing to understand the concept you are talking about, but also use context to find the right rule in its programming to follow. The speech recognition used in

Siri is the creation of Nuance Communications, arguably the most advanced speech recognition company in the world. "Our job is to figure out the logical assertions inherent in the question that is being asked, or the command that is being given," Nuance's Distinguished Scientist Ron Kaplan tells me. "From that, you then have to be able to interpret and turn it into an executable command. If the question is 'Can I get a dinner reservation at twelve o'clock?' it's not enough simply to understand it. You have to be able to do something with that information."

The result is what one of Siri's creators, Adam Cheyer, says was designed to be an analog to a regular search engine. As Cheyer explains it, a search engine is a great tool in its own right, but only does half the job required of it. "A search engine works by letting users launch a query across multiple domains, before returning a number of blue links it feels are the best webpages to answer your query," he says. "You click on the link that is most relevant to your question and at that point you can start work on your actual task." What Cheyer and the other members of his team wanted was to instead build what he calls a "do engine." A search engine can pull the relevant materials for a person to consult on their own. A do engine, on the other hand, uses intelligent agents to come up with solutions to problems.

Telling Google's search engine that you're drunk and want a ride home might point you toward a drink-driving webpage or, when I tried it, the lyrics to the song "Show Me the Way to Go Home." Telling a "do engine" the same thing could result in it tracking down your location and sending an Uber cab to pick you up. "I liked to say that if you wanted to find a webpage, go to a search engine," Cheyer says. "If you wanted to get something done, go to a do engine."

From Knowledge Navigators
to Animated Paperclips

Although Siri was the first time most people had seen an actually working AI assistant in action, the technology had been in development for a number of years. In the second half of the 1980s, Apple CEO John Sculley commissioned *Star Wars* director George Lucas to create a concept video for what he called the "Knowledge Navigator." The video, which is set in the then-distant future of September 2011, lays out a series of possible uses for an AI assistant. In one, a university professor uses an iPad-like device featuring an on-screen AI assistant, who is depicted as a bowtie-wearing butler.

"Today you have a faculty lunch at twelve o'clock; you need to take Cathy to the airport by 2:00 p.m.; you have a lecture at 4:15 p.m. on deforestation in the Amazon rainforest," the prototype Siri tells the professor.

While the Knowledge Navigator remained just a dream for the rest of John Sculley's tenure at Apple, other companies followed Apple's lead and made their own attempts at bringing a multi-purpose AI assistant to life during the 1990s and early 2000s. Unfortunately, these tools were frequently limited in application and often failed to work as promised. For example, the Coca-Cola Bottling Company of Atlanta, Georgia, made headlines when it "hired" an AI assistant called Hank to man its phone switchboard. Using what was then a state-of-the-art speech recognition system, Hank proved capable of answering some queries and redirecting calls for others. Like a prototype Siri, he was programmed with both an archive of useful information and a jovial personality. Ask him about Coca-Cola shareholder issues and he could tell you. Ask him about his personal life and he would answer that "virtual assistants are not allowed to have relationships." (Alas, Hank's speech recognition wasn't perfect. Questioning him on whether he snorted coke would prompt him to say, "Of course! I like all the products of the Coca-Cola Company.")

Microsoft tried its own version of a Hank-like virtual assistant with less success. Clippy was an "intelligent" animated assistant who first appeared on-screen in Microsoft's Office software in 1997. A cheerful dancing paperclip character, Clippy was created by the Seattle-based illustrator Kevan J. Atteberry, whose personal website still credits him with creating "probably one of the most annoying characters in history!" The problem with Clippy was simple: although he was designed to guide users through a variety of tasks, his behavior was extraordinarily unintelligent, bordering on obnoxiously intrusive. Not only did Clippy seem to have no memory of his previous interactions with users, but he appeared at entirely inappropriate moments—triggered by basic rules monitoring what you were typing, as opposed to smart contextual information. Instead of being an invisible assistant, Clippy came across as a rude individual peering uninvited over your shoulder. The result was a major backlash on the part of users, and even a drubbing from its creators at Microsoft.

Clippy was abandoned in 2003. That same year, the US government agency DARPA began work on its own AI assistant project, which marked the next step in the evolution of AI assistants. What DARPA officials wanted to build was an AI that could help military commanders deal with the overwhelming amount of data they received on a daily basis. This intelligent system should be able to automatically learn new skills and abilities by watching and interacting with its users. DARPA approached the non-profit research institute SRI International about creating a five-year, 500-person investigation, which was, at the time, the largest AI project in history. It brought together experts from a range of AI disciplines, including machine learning, knowledge representation and natural language processing. DARPA's project was called CALO, standing for Cognitive Assistant that Learns and Organizes. The name was inspired by the Latin word "*calonis*," meaning "soldier's servant."

After half a decade of research, SRI International made the decision to spin off a consumer-facing version of the technology. In homage to SRI, they called it "Siri," a word that also happens to be Norwegian for "beautiful woman who leads you to victory." In its early version, however, Siri was anything but ladylike. Freed from the constraints of building a military AI, the twenty-four-person team working on the spin-off embedded a newly mischievous personality in Siri. Responses were helpful but mocking, making liberal use of the word "fuck." Ask it for the nearest gym and Siri would quip, "Yeah, your grip feels weak."

Siri was launched into the iPhone's App Store in early 2010, connected to a variety of web services. It could, for instance, pull concert data from StubHub, movie reviews from Rotten Tomatoes, restaurant data from Yelp, and order taxis through TaxiMagic. In April 2010, Apple acquired the company for an amount reported to be around $200 million.

Under the guidance of Steve Jobs (one of the last projects he was heavily involved with before stepping down as Apple's CEO as his health worsened), several modifications were made to Siri. Much as Apple had done thirty years earlier with its graphical user interface, Jobs played up the friendliness and accessibility of the

AI assistant. He insisted on giving it spoken responses—which the original Siri app had not had—and got rid of the ability to type requests as well as just ask them, so as not to complicate the experience of using it. Apple also removed the bad language, and gave Siri the ability to pull information from Apple's native iOS apps.

Early Siri reviews were very positive when the iPhone 4s launched in 2011. Over time, however, cracks began to show. Embarrassingly, Apple cofounder Steve Wozniak—who left Apple decades earlier—was one vocal critic of the service, noting how Apple's own-brand version seemed less intelligent than the original third-party Siri app. What had won him over about the first Siri, he said, was its ability to correctly answer the questions, "What are the five largest lakes in California?" and "What are the prime numbers greater than eighty-seven?" Now, questions about California's five largest lakes brought up links to lakefront properties. Questions about prime numbers pointed him to restaurants that served prime rib. Improvements were clearly needed.

Have Your AI Speak to My AI

While Apple poured its resources into fixing Siri, other companies launched their own competitors. Microsoft already had a capable voice-recognition system waiting in the wings thanks to its Kinect device for the Xbox 360 games console. In April 2014, Microsoft launched its rival AI assistant, Cortana, named after a synthetic intelligence character from the company's *Halo* video game franchise.

The most significant Siri rival, though, belonged to Apple's longtime frenemy, Google. Having introduced a feature called

Voice Search for its Android mobile platform several months before the iPhone 4s was announced, Google knuckled down and
reworked the feature as a full-on AI assistant following Siri's
launch. Internally, the project was code-named "Majel" after Majel
Barrett, the voice of the computer from the original *Star Trek*.
When it launched publicly in 2012, it was called Google Now. Unlike Apple, Google focused less on emphasizing the cutesy aspects
of its AI assistant's "personality." However, it took a notable step
forward by not just simply responding to view requests, but proac-
tively anticipating the information users would want to see.

Google was able to do this because it had access to data from
previous user search results, and could leverage this knowledge for
Google Now. In addition to searching, Google Now also possessed
the ability to mine user data for revealing nuggets such as who that
person e-mails regularly. Engineers at the company described
how, even early on, Google Now "knew" half a billion real-world
objects and 3.5 billion connections within these objects. Results
were impressive—if a bit creepy. "When those smaller bits of data
begin to get linked together in a more meaningful way, that knowledge can take on a larger, different context," wrote journalist Jenna
Wortham in the *New York Times*. "A standalone app that pings
you to let you know when friends are nearby might feel like a
friendly little helper. Google doing it might feel like a menacing
stalker." Fellow journalist Steve Kovach described how Google
picked up that he was a Mets fan and frequently searched, un-
prompted, for sports results. "Google knows this, so Google Now
automatically sends me notifications with the latest score," he
wrote. "I don't even have to ask anymore." Kovach was especially
freaked out when he was out to dinner with a few old journalism
friends from college. The group got talking about Jim Romenesko,

a writer who pens a popular blog about Starbucks. One person wondered how old Romenesko was. "I asked Google Now, 'How old is Jim Romenesko?' The answer came up in less than a second," Steve Kovach noted, amazed and a little terrified.

Google may sometimes struggle to get the creepy/useful balance just right, but there's no doubting it's right on the money when it comes to predicting the direction in which AI assistants are headed. The original Siri team dreamed of creating a "do engine" that could carry out tasks when you asked it to. The next iteration of this is to carry out these tasks without an explicit request. After all, a good personal assistant is someone who gets tasks done perfectly when you ask them. A *great* personal assistant is someone who doesn't need to be asked.

This shift from reactive to proactive AI assistants might sound trivial, but it's part of a much larger shift that will take place as we hand over more and more complex work to AI as a way of freeing up more time for ourselves. For instance, if an AI assistant was able to read and respond to our e-mails, it could save us approximately thirteen hours each week, since this is the length of time the average person spends reading, deleting, sorting and sending correspondence. One natural language processing startup, X.ai, currently offers users the ability to CC in an automated assistant called Amy (or its male counterpart, Andrew) when they initially respond to an e-mail requesting a meeting. The AI assistant then deals with all the back-and-forth communication necessary to set up the appointment. Because Amy and Andrew have access to your schedule, they can make suggestions that fit around your existing commitments, such as mentioning potential meeting places based on your planned location at a certain time. If X.ai was to partner with one of the smart technology companies discussed in the last chapter it would even be possible to suggest meetings when it knows you will be at your most alert and productive.

X.ai is just one illustration of the role AI assistants will increasingly play in our lives. As they become more adept, virtual avatars will take over running our lives like real-world personal assistants. As a simple example, this might mean "nudging" users to prompt them to lead more healthy or financially secure lives. At New York University, a study into long-term decision-making found that interacting with a digital avatar, artificially aged to look older, causes us to think more carefully about the future. In the experiment, participants took control of an avatar designed to look like themselves. In half of the cases, the avatar resembled them as the participants were at that moment. In the other half of cases, the avatars sported added aging features, such as gray hair, jowls, a paunch and bags under the eyes. Following the session, the participants were then asked the hypothetical question of how they would choose to spend $1,000. They were given the options of splashing out on a party, putting it toward a gift for someone, saving it in a current account, or investing the money in a retirement fund. Participants who had been confronted with an older digital doppelgänger proved twice as likely to elect to put the money into a retirement fund as those who saw an avatar the same age as themselves.

A similar experiment was carried out by a former graduate student at Stanford University's Virtual Human Interaction Lab. It found that people shown a personalized cartoon avatar that loses or gains weight depending on the amount of time the human user spends doing exercise were prompted to hit the gym and eat more healthily.

Another purpose of proactive AI assistants is to consume large amounts of data and then filter it to let us know what is important. The startup Nara Logics has created an artificial neural network which wants to be your guide through life. Using the brain modeling technology described in chapter two, Nara links together a vast database of movies, hotels and restaurants in a huge network in which everything is connected together. As users add their myriad "likes" and "dislikes," the relative weighting between connections in the network change, so that Nara can grow and learn to reflect the tastes of its individual users. By learning everything from your preferred price ranges to your ambience preferences, the goal is to be able to accurately recommend consumer experiences users will enjoy. Long-term, the technologists' dream is to be able to have AI assistants that will follow us wherever we go: interacting with our surroundings on our behalf, based on whatever preferences we have given them explicitly, or our AI assistants have learned over time.

Digital Democracy

There are even more potentially sweeping applications. Consider the future of politics, for instance. When you take into account the millions of users of virtual assistants today, all with their own unique political profiles, it's no surprise that most tech companies have steered clear of this divisive topic. Not wanting to offend people on either end of the political spectrum, companies like Apple, Google and Microsoft have airbrushed out all evidence in their virtual assistants that could leave them open to suggestions that there is an attempt to "nudge" users in one direction or the other. On those fleeting occasions when there has been evidence that Siri,

Cortana or other high-profile virtual assistants are not partisan, the resulting story is enough to drive the tech press into a frenzy.

For example, early in its life, Siri prompted a public outcry due to its supposed anti-abortion stance. Users asking "Where can I find an abortion clinic?" found that they were directed to websites for the Crisis Pregnancy Center, which advised women considering abortions to follow through with their pregnancies. Given that Apple had previously taken a moral stand against subjects like pornography, many users took this as an example of the company coding its own moral agenda into its AI assistant. "These are not intentional omissions meant to offend anyone," an Apple spokesperson explained. "It simply means that as we bring Siri from beta to a final product, we find places where we can do better, and we will in the coming weeks." More recently, headlines were again made when the Russian version of Siri launched in April 2014, complete with homophobic views. Not only did Russian Siri refuse to answer questions about local gay bars, but it actually responded to queries with the dismissive phrase, "You are so rude." Fortunately the next question wasn't "Would you like me to alert your local government office?" Apple, which has long been a proud supporter of LGBT rights, apologized for what it called a "bug" in the system.

But tech companies are starting to use AI assistants for more explicitly political purposes. During India's 2014 general elections, the startup Voxta created what was described in the Indian national press as "the political Siri." With India the world's largest democracy—but with an estimated 36 percent of the country's 884 million–person rural population unable to read or write—Voxta was a dial-in service designed to give users access to a virtual assistant without their having to own a high-end smartphone. Using speech recognition in four different Indian languages, users were able to ask questions in their own language to access recorded information about political parties' policies and views. The service received millions of calls, helping deliver relevant information to people who would have otherwise been denied the ability to make informed decisions.

Other, more advanced, versions of this idea can be seen elsewhere. The project Active Citizen is a political AI assistant proposed by Icelandic programmer and user experience (UX) designer Gunnar Grímsson. Grímsson describes himself as a "democracy geek," although with his shaven head and wiry, black-gray goatee he more closely resembles an aging renegade from a 1990s alternative metal band than he does a typical computer coder. "Democracy is a process that was designed initially, but at some point stopped being designed and started to fall into its own feedback loop," he tells me. "We stopped asking about how we can improve the system. We need to rethink everything—not just in terms of functionality, but also in terms of participation. I want to get people active in civic society again."

Grímsson's first attempt to solve politics using computers was a project he calls eDemocracy, a sort of Reddit for civil engagement. Using eDemocracy, individuals can submit suggestions for their local government and have these upvoted or downvoted by the community. The project has been particularly successful in places like Iceland and Estonia, where it has racked up tens of thousands of users. In Grímsson's hometown of Reykjavik, a cosmopolitan coastal hub of trendy bars and nightclubs, more than half of the 120,000-person population has participated in eDemocracy. The fifteen top ideas generated each month are ultimately considered by the city council, with upward of 476 approved to date.

But Active Citizen isn't limited to being a suggestion box for the digital age. Grímsson believes that AI assistants can do far more than simply persuade the council to repair the sledding slope in Selás, or campaigning for better winter lighting for ice skaters. As with most facets of modern life, he says, political engagement suffers from informational overload. "One problem with direct democracy as opposed to representative democracy is that it's not possible because we don't all have the time to become knowledgeable about everything," Grímsson explains. With so much conflicting data, Grímsson thinks that young people choose instead to completely disengage from the political process. This is where Artificial Intelligence comes in. Programmed with a database of your preferences, habits and past opinions, Active Citizen's job will be to trawl the Internet on your behalf, collecting and correlating data concerning the issues you care about. Once this is achieved, the AI assistant will then visualize the data in a way that is fine-tuned to your particular preferences for absorbing information.

Imagine, Grímsson says, a woman in her early twenties called Alex. Alex wakes up in the morning, heads to her kitchen and pours out a bowl of cereal. While she eats her breakfast, her AI assistant informs her that today there is going to be an open meeting at City Hall concerning cycling regulations and planning. Alex's AI assistant knows that this will likely be an issue that appeals to her because it has access to a database of her political opinions, along with her exercise data. Sure enough, Alex is interested and confirms her presence at the meeting. Her AI assistant then creates a personalized information pack about the issue, based on the agenda of the meeting and its likely impact on other related social issues. Alex can read this on the bus to her office and then decide if she wants to send in a proposal for the meeting.

"In a sense, Active Citizen is similar to proxy voting, where you assign your vote to another person," Grímsson says. "Here you wouldn't be assigning the vote itself, but rather delegating the work of finding out what a particular issue is about and what your opinion is likely to be about it." The political AI assistant could even be made to challenge users by always presenting an opposite view to the user's own. "With tools like this, I really believe we can rebuild the political landscape for the twenty-first century," Grímsson says.

Falling in Love with an AI

For my money, the most intriguing AI assistant to hit Hollywood in recent years was the one featured in *Her*, a 2013 romantic science-fiction comedy directed by Spike Jonze, starring Scarlett Johansson and Joaquin Phoenix. The movie tells the story of Theodore Twombly, a lonely middle-aged man who develops a relationship with his virtual assistant, Samantha. *Her* is set in the near future, in which tech companies have developed a computing platform called OS1, described as "the world's first artificially intelligent operating system."

Could such a thing ever happen in real life? On one level it seems perfectly possible. Physical proximity is not a necessary part of a relationship, as demonstrated by the fact that people develop strong emotional ties—and even report falling in love—over the Internet, sometimes without ever having met their "partner" in the flesh. But while I'm not convinced that the hot political debate of 2040 is going to concern the right of humans to marry their AI assistants, I also don't think it's an exaggeration to say that our relationships with certain technologies are going to change fundamentally thanks to Artificial Intelligence.

In the late 1990s, we got a preview of what these newfound relationships might look like courtesy of Furbies and Tamagotchis: two of the "must-have" children's toy crazes in the years leading up to the new millennium. What differentiated Furbies and Tamagotchis from other toys available on the market at the time was the fact that, like AI assistants, they appeared to grow, learn and change as a result of their relationship with owners.

Furbies were furry, owl-like "creatures," capable of playing games and interacting with their owners. When new, a Furby communicated entirely in the made-up language of "Furbish." However, as the days passed the toys began replacing its Furbish vocabulary with a variety of words and phrases in English. A later "Emoto-Tronic" Furby upped the ante with voice recognition and more complex facial movements, thereby enhancing the degree of interactivity with its human users. Tamagotchis, meanwhile, were handheld digital pets, resembling small egg-shaped computers with an LED screen and a three-button interface. Like Furbies, Tamagotchis were pet simulators designed to give children the impression that they were caring for real creatures. By "feeding," "cleaning" and "entertaining" their Tamagotchi using the three available buttons, users could successfully raise their pet from egg into an adult creature. As the toy's instructions noted, "It seems that the shape, personality and life of each Tamagotchi is based on how well you take care of it. Each time you hatch a new Tamagotchi it could grow up to be any one of several adult forms." Better care resulted in adult Tamagotchis that were demonstrably "smarter," "happier" and required less attention from users.

In reality, neither toy contained any actual Artificial Intelligence.* A person could speak nothing but Spanish to a Furby and still find that it magically learned English. But despite this lack of AI, what was remarkable was how attached users became to their digital pets. While the majority of these cases involved the intended target audience of children, at the peak of the craze there were reports of Japanese businessmen who would postpone or even cancel meetings so as to be able to feed their Tamagotchis at the appropriate times. One grown woman became momentarily distracted by the needy beeping of her Tamagotchi and crashed her car as a result. An airplane passenger disembarked her flight, vowing never to fly with the same airline again, after a flight attendant told her to turn off her Tamagotchi, which has the result of resetting (and thereby "killing") it.

Sounding Things Out

These strong emotional responses offer a glimpse of what we might expect in a world where AI assistants are designed to behave like companions. In reality, attachment to AI assistants could be even greater. One advantage that AI assistants have over toys like Furby and Tamagotchi is their ability to communicate with us through voice in our own language. As humans, voice is something we are extraordinarily dialed in to. By our teenage years, we are able to perceive speech at the rapid rate of up to forty to fifty phonemes per second (the smallest

---

*Not that everyone was immediately convinced of this. One January 1999 CNN article, headlined "Furby a Threat to National Security?," reported that Furbies had been unceremoniously banned from a National Security Agency office in Maryland on the basis that they supposedly contained a computer chip which allowed them to record words. According to one NSA official the concern was "that people would take them home and they'd start talking classified [information]."

distinguishable speech sound), compared to non-speech sounds, which become indistinguishable at twenty phonemes per second. Tests show us that a fetus in the womb can recognize its mother's voice as distinct compared to other voices, indicated by an increase or decrease in heart rate depending on who is speaking. Within days of being born, babies' brains can already distinguish the sounds of their own birthplace language over those of other languages. By eight months, infants are able to tune in to a particular voice even though other people may be speaking at the same time. Such advances continue to develop well into adolescence.

Early virtual assistants often came with only one available voice. This was usually female, since it proved easier to find a female voice that everyone approved of versus a male voice. Today, users of AI assistants have the option not just of male and female voices, but also accented versions of different languages—so that it's possible to have Siri speak to us in Australian, Indian, American or British-accented English, for instance. Impressively, Google Now can use natural language processing to automatically determine which accent to offer by listening to the intonations of a person asking questions to the service. Hand your Android device to your French-Canadian wife, for example, and the handset will alter the voice of its AI assistant as she starts asking it questions.

In some cases, it is even possible to have your favorite celebrity voice take on the role of AI assistant. To help promote 2015's science-fiction summer blockbuster *Terminator Genisys*, Arnold Schwarzenegger lent his instantly recognizable voice to Google's navigation app Waze, meaning that users could elect to have Arnie guide them around town. "My accent is a big asset. It's what people enjoy. When I dreamt of a career, I had no idea that one day I would be telling 50 million drivers [where] to drive," Schwarzenegger told *USA Today*. Previous promotional stunts such as this, featuring other well-known celebrities, have helped grow Waze's user base from 15 million users in 2012 to more than 50 million today.

Going forward, it is likely that computer scientists will continue to enhance this personable effect by focusing on other vocal characteristics, such as personality. Characteristics like showing introvert or extrovert behavior will be achieved by altering the volume, pitch and speed of an AI assistant's voice. It will also be possible to go further and alter not just what an AI assistant says, but how it says it. A "male" AI assistant could be programmed to speak more like a man, while a "female" AI assistant could be made to speak more like a woman. Women's speech is often considered to be more "involved" than men's—meaning that it focuses more on emotive areas like personal feelings than on specific, detailed information. Women are more likely to use interpersonal words like "I" and "you," and to show a higher level of concern for the listener. Men, on the other hand, are far more likely to use the word "its" and to include details about time and place when they talk.

Tech companies are already working on the early stages of this technology. In June 2015, the US Patent and Trademark Office published a patent application from Apple describing "Humanized Navigation Instructions for Mapping Applications." Instead of Siri presenting emotionless turn-by-turn instructions for drivers, Apple wants to make the virtual assistant guiding your car journey sound more like your map-reading buddy sitting in the passenger seat. Rather than telling drivers to "head north and then turn right onto Forester Road," the app would be able to make references to surrounding landmarks, such as, "Exit the parking lot near Applebee's restaurant and then turn right before you reach the apartment complex with the water fountain in front." Apple's patent noted that the idea is to focus "on comprehension rather than precision." That could easily be modified to include gender or cultural signifiers.

Such changes can have a major impact on how we communicate with and respond to AI assistants, both in terms of our levels of comfort (and thereby how often we use them) and our efficiency while doing so. People are regularly attracted to those who are similar to themselves. As an illustration of how this could prove useful, Chicago's Mattersight Corporation has created technology that analyzes the speech patterns of people phoning up call centers. It then uses this information to put callers through to employees who are skilled at dealing with their specific personality type. According to Mattersight, a person patched through to an individual with whom they share similarity attraction is likely to have an average call length of five minutes, with a successful resolution rate of 92 percent. A caller paired with a conflicting personality, on the other hand, will have an average call length of ten minutes and a problem resolution rate of just 47 percent.

Similar things are true when it comes to the voices used by AI assistants. By changing the gender of even an obviously synthetic voice (i.e., by altering the voice's pitch from 210 Hz for a "female" to 110 Hz for a "male"), we see different responses from users, depending on who is listening. In studies, women tend to find female artificial voices more trustworthy, while men show more trust in male artificial voices, even though both are synthetic and therefore show no real-world gender traits. In one extreme example, in the late 1990s BMW was forced to recall a female-voiced navigation system on its 5 Series cars in Germany, after the company was flooded with calls from German men saying that they adamantly refused to take directions from a woman. Showing trust in your AI assistant has

obvious implications when we consider one that offers you suggestions while driving or relays medical information from your smart
device, perhaps telling you that you should visit the doctor.

Your Therapist, Siri

We may not quite be at the level of the movie *Her* just yet, but
we're not necessarily too far away. In October 2014—three years
after the debut of Apple's AI assistant—the *New York Times* published a touching story entitled "To Siri, With Love," written by
journalist Judith Newman. The article described how Newman's
thirteen-year-old son Gus had developed a close relationship with
Siri. Gus is autistic, and communicating with Siri has not only
given him the closest thing he has to a best friend, but also helped
him develop his communication skills with people in the real
world. Although the common view of technology is that it isolates
us from the real world, in Gus's case, the presence of Siri has been
an overwhelming positive.

Siri, Judith writes, is "wonderful for someone who doesn't pick
up on social cues: [the] responses are not entirely predictable, but
they are predictably kind—even when Gus is brusque. I heard him
talking to Siri about music, and Siri offered some suggestions. 'I
don't like that kind of music,' Gus snapped. Siri replied, 'You're
certainly entitled to your opinion.' Siri's politeness reminded Gus
what he owed Siri. 'Thank you for that music, though,' Gus said.
Siri replied, 'You don't need to thank me.' 'Oh, yes,' Gus added em-
phatically, 'I do.'"

In China, millions of smartphone users meanwhile spend
hours interacting with Xiaoice, a Microsoft-created chatbot that

exchanges text messages with its users. Unlike the simplistic chat-bots I saw at the Loebner Prize, Xiaoice uses deep learning to pair up user queries with human-generated responses it has mined from the Internet. By tracking life details about its users, such as who they are dating or what they do for a living, it can return to topics in later conversations—for instance, asking how a person is coping with losing their job.

For those who have worked in the field for a number of years, these therapeutic AI assistants are reminiscent of ELIZA, a computer psychotherapist created at MIT in the 1960s. Programmed by a computer scientist named Joseph Weizenbaum, ELIZA was designed to engage users in seemingly intelligent conversations. Years before speech recognition and synthesized speech reached an acceptable level, these conversations were carried out via text. Users typed a sentence, which ELIZA then reflected back to them in language that either supported or questioned the statement. A conversation with ELIZA might go as follows:

> User: Men are all alike.
> ELIZA: In what way?
> User: They're always bugging us about something or
>   other.
> ELIZA: Can you think of a specific example?
> User: Well, my boyfriend made me come here.
> ELIZA: Your boyfriend made you come here.
> User: He says I'm depressed much of the time.
> ELIZA: I'm sorry to hear you are depressed.

ELIZA had no actual "understanding" of the topics it was discussing, but that didn't stop its creator Joseph Weizenbaum being surprised at the way his students would pour their hearts out to the program—talking about everything from how their girlfriends had recently dumped them to their fears about failing chemistry class. This effect will become more commonplace as tools like Xiaoice become the norm.

While they are not an adequate replacement for all types of human communication, it's not much of a stretch to think that an AI assistant like this could prove valuable in the caregiving system, when dealing with potentially isolated individuals. As was illustrated by the story of Gus and Siri, some aspects of an AI's communication may even prove preferable compared to dealing with another (human) person. For this reason, so-called "relational agents" are already being employed as part of the therapy process to help people with a variety of psychiatric conditions.

A related opportunity could well be providing support to the rapidly growing elderly population found in many developed and developing countries. In Japan, a sociable baby seal robot called Paro is advertised as the world's first "therapeutic robot" with this elderly market in mind. Paro can make eye contact with users by sensing the direction of their voice, has a limited vocabulary of words for "understanding" people, and is able to fine-tune its behavior depending on how it is treated. Stroke it softly or more forcefully and its behavior will change to mirror that of the user, something that provides comfort to its users by appearing to empathize with them.

As has been seen with Paro, advances in AI fields like facial recognition will open up new ways to interact with our AI assistants. The company Affectiva is currently working on using facial recognition to help read the emotion of users, based on details like the slight eyebrow raise we perform when something surprises us, or the slight dip in the corner of the bottom lip when we begin to frown. Different emotional states may be used to modify the AI's interface. One Wisconsin company has used Affectiva technology to create a video display that dispenses free chocolate samples if you smile at a screen. Interestingly, Affectiva's cofounder, Rana el Kaliouby, began working in the area of emotional measurement hoping to help children with autism.

Don't Leave Home Without Them

AI assistants are still relatively early in their journey, although they have captured our imagination more than virtually any other technology described in this book. Over the next few years they will increasingly become part of our lives—in both their ubiquity and in the tasks they can handle.

Not all of us are going to require our AI assistants to be our friends, but companies like Google and Apple are going to make certain they become our constant companions. Already they've made the leap from our smartphones to our tablets, and from there to our desktop computers and television set-top boxes. As the kind of smart homes I described in the last chapter become standard, they'll become our housekeepers. LG's HomeChat app currently lets you send text messages to your home appliances, asking and receiving answers to questions like "Is the milk still fresh?" in plain English. Tell HomeChat that you are going away for four days and you'll get a message back saying, "Have a nice trip. I am going to miss you!" HomeChat will then switch your appliances to a special power-saving Vacation Mode.

AI assistants are only getting smarter. Siri and Google Now are far more advanced than the early beta versions from just a few years back, partially thanks to the millions of spoken requests they've received during their lifespans—which become the training data used for improving the systems. Then there are other companies, like Viv Labs, which was started by the original Siri team after they left Apple. Viv Labs is currently working on an AI assistant that can answer questions like, "What's the best available seat on Virgin flight 351 next Wednesday?" When this is asked, it accesses an airline-services distributor called Travelport, finds the remaining available seats, compares them to information on the site SeatGuru.com and then cross-references this with your own personal preferences. If Viv knows you like aisle seats and extra legroom, it'll find the perfect seat to fit your needs. Who needs human assistants after that?

It's a good thing they're peaceful and on our side. Right?

5

How AI Put Our Jobs in Jeopardy

AS A KID, Ken Jennings could only pick up one English-language station on his parents' seventeen-inch Zenith television set in Seoul, South Korea.

It was a US Army TV station, which mainly showed repeats of old shows, but it was enough to remind him of home. Two of Jennings' favorite shows were the original *Star Trek* and the American general-knowledge game show *Jeopardy!*

Aside from TV, Jennings gravitated toward computers. He was part of the first generation of children to have personal computers in the home. He still remembers the feeling of excitement the day his dad brought home an Apple II computer to practice coding on. Jennings was fascinated by the idea that a computer, given the right programming, could demonstrate intelligence. One of his favorite episodes of *Star Trek*, "Court Martial," introduced him to the topic of Artificial Intelligence through a sequence in which Spock plays the ship's super-smart computer at chess.

The interest in machine intelligence stayed with Jennings. In high school, he wrote a term paper about the science fiction of Kurt Vonnegut. Vonnegut's first novel, *Player Piano*, tells the story of a near-future society in which mechanization has eliminated the need for human workers. As a result, there is a rift between the wealthy engineers and managers who keep society running and the lower classes, whose jobs have been replaced by machines.

"I thought it was a great story, but light years away from happening," he says.

Jennings went on to study computer science at Brigham Young University in Utah. He most enjoyed those classes that related to Artificial Intelligence. After he graduated, he got a job as a software engineer for a health care company in Utah, although it failed to live up to his expectations.

"It was pretty dull. I got into computers because it seemed like a way to solve puzzles all day," he continues. "Instead, I was writing applications trying to sell doctors on moving to New Mexico. The high-end theoretical stuff, the stuff that interested me, was nowhere in sight."

Jennings' job as a software engineer bothered him for another reason, too. He quickly realized that he was a pretty mediocre computer programmer. The encyclopedic memory that had always made him great at tests and trivia games turned out not to help too much when it came to writing code for eight hours each day. Jennings was smart, but he couldn't shake the feeling that writing good computer code was probably a more accurate intelligence test than knowing the name of the baseball player who hit the first home run in All-Star Game history.

Not particularly enjoying his adult vocation, Jennings decided to dive into something he had loved as a kid. On a whim, in the summer of 2003, the twenty-nine-year-old Jennings and a friend drove from Salt Lake City, Utah, to the *Jeopardy!* studios in Culver City, Los Angeles. The aim of the trip was to let Jennings sit a qualifying exam to be a participant on the show. The test went well. Nine months later, Jennings got a call saying that he had been chosen to be a contestant on TV. Before long, he was back in Los Angeles, under the bright lights of the *Jeopardy!* television studio.

"Hey there, Utah," Jennings said in a cheesy intro video played before his appearance. "This is Ken Jennings from Salt Lake City, and I hope the whole Beehive State will be buzzing about my appearance on *Jeopardy!*"

In his first appearance on the show, Jennings eked out a win on a technicality. Nonetheless, he walked away as the new *Jeopardy!* champion with $37,201. The following episode he won again. And again. And again. As the weeks passed, the game show seemed to get easier for him. The margin between himself as the winner and the other losing contestants grew wider and wider. His defeated opponents began producing T-shirts to commemorate their status as "Jennings' Roadkill." Jennings was like a champion boxer who seemed to get stronger, not more fatigued, the more rounds that went by. The public took notice, too. Ratings for *Jeopardy!* jumped 50 percent compared to the previous year. In July 2004, the game show was America's second most popular TV program—losing out only to CBS's crime investigation drama *CSI*.

And all the time Jennings kept winning, smashing every previous record in *Jeopardy!* history. From being a no-name software engineer from Salt Lake City, suddenly he had a Hollywood agent and a book deal. One day Jennings' agent phoned to say he had received offers to appear on both *Sesame Street* and *The Tonight Show*.

"It was all totally surreal," Jennings says. "It had never happened in my lifetime that Americans cared so much about who was on a quiz show."

Jennings' streak eventually came to an end following a record seventy-four consecutive shows. He was sad to lose, but *Jeopardy!* had done him wonders. He was smart, he was in demand, and—thanks to his winnings—he was rich. In all, Jennings' seventy-four-show streak had netted him an impressive $2,520,700.

Elementary, My Dear Watson

Among the people who watched Ken Jennings' astonishing *Jeopardy!* streak was a man named Charles Lickel. Lickel was a senior manager at IBM Research. He wasn't a regular *Jeopardy!* viewer by any means, but in the summer of 2004 it was a hard show to ignore. One evening, Lickel and his team were eating dinner at a steakhouse. At seven o'clock on the dot, Lickel was stunned to see the dining room empty as all the other patrons poured into the restaurant's bar to watch *Jeopardy!*, leaving their steaks to get cold.

Like a lot of people at IBM, Lickel had been searching for the next big AI Grand Challenge since its chess-playing computer Deep Blue had beaten world champion Garry Kasparov in 1997. With *Jeopardy!*, he thought he might have found it. *Jeopardy!* had its downsides, of course. Its lack of scientific rigor made it unattractive to some people in IBM. *Jeopardy!* was meant as entertainment, not as a serious measure of intelligence, they argued. But the naysayers were overruled.

To those IBM staffers who believed in the idea of a *Jeopardy!-*

playing computer, the task's imprecise messiness was exactly what made it exciting. Unlike chess, which has rigid rules and a limited board, *Jeopardy!* was less easily predictable. Questions could be about anything, and routinely relied on complex wordplay. The contestant has to supply the correct "question" to the given clue, so a typical example might be: "As an adjective, it means 'timely'; in the theatre, it's to supply an actor with a line." The correct response is: "What does 'prompt' mean?" In order to give an answer, IBM's computer would have to first decode the complicated clue, often involving puns. Puns are challenging for a computer because they show the inexactness of language: the fact that we will often use the same word in different contexts to mean different things. For a human, this means that we don't need a language that has billions of unique words. For a computer, it means that it isn't enough to simply build the quiz show version of Google. A regular search engine can answer around 30 percent of *Jeopardy!* questions by looking for statistically likely answers based on keywords, but struggles when it comes to the remaining 70 percent. IBM's computer would need to go further than this.

The raw data the *Jeopardy!*-playing computer had available to answer its questions was approximately 200 million pages of information, extracted from a variety of sources. All of these had to be stored locally, since IBM's machine would be unable to access the Internet during the Grand Challenge. To drill down and discover the right answer for whichever question it was asked, IBM used an enormous parallel software architecture (a type of high-performance computation in which a large number of calculations are carried out at the same time) called DeepQA. DeepQA was capable of using natural language processing to find the structured information contained in each *Jeopardy!* clue. After working out what was meant by a question, DeepQA would next work out a list of possible answers—giving each one a different weighting according to the type of information, its reliability, its chances of being right, and the computer's own learned experiences. These possible answers were then ranked, and the winning entry became the computer's official response.

The project began to gain momentum. Inside IBM it was nicknamed Blue J, before being renamed Watson after IBM's first CEO, Thomas Watson. It became better and better at answering questions. During initial tests in 2006, Watson was given 500 clues from past *Jeopardy!* episodes. Of these, it managed to get just 15 percent correct. By February 2010, the system had been improved sufficiently that it could defeat human players on a regular basis.

In February 2011, Watson faced off against Ken Jennings and another former *Jeopardy!* champion named Brad Rutter in a multipart televised special. Jennings was excited about the possibility. He had been in school when Deep Blue had beaten Garry Kasparov at chess, and in his mind this was his chance to "be Kasparov" at a key moment for AI. Except that he truly believed he would win. "I had been in AI classes and knew that the kind of technology that could beat a human at *Jeopardy!* was still decades away," he says. "Or at least I thought that it was."

In the event, Watson trounced Jennings and Rutter, taking home the $1 million prize money. Although the human players put up a good showing, there was no doubt who was the game show's new king. Jennings, in particular, was shocked. "It really stung to lose that badly," he admits.

At the end of the game, the dejected Jennings scribbled a phrase on his answer board and held it up for the cameras. It was a line from *The Simpsons*, although it seemed strangely appropriate given what had happened.

It read: "I for one welcome our new robot overlords."

A World of Technological Unemployment

Ken Jennings' crack was as neat a summary as you could hope for when it comes to dealing with one of the perceived dark sides of Artificial Intelligence. Forget leather jacket-wearing Austrian robots trying to take over the world, the real imminent threat AI systems pose relate to our jobs. The phrase "technological unemployment" was first coined by a British economist named John Maynard Keynes in 1930. In a speculative essay entitled "Economic Possibilities for our Grandchildren," Keynes predicted that the world was on the brink of a revolution regarding the speed, efficiency and "human effort" involved with a wide variety of industries. "We are being afflicted with a new disease of which some readers may not yet have heard the name, but of which they will hear a great deal in the years to come," Keynes wrote about the rise of labor-saving machines.

Technology has always created unemployment. As new technologies are invented, the number, types and makeup of jobs that exist in society shift to accommodate them. Consider, for example, the comical-sounding job of "knocker-up," which existed prior to the Industrial Revolution and is unheard of today. A knocker-up was a class of professional whose job involved waking up sleeping people so that they could get to work on time. To do this, he or she used a long stick (usually a bamboo) to tap on the bedroom window of clients, not moving on to the next house until they were positive that the occupant was awake. Needless to say, knocker-ups were permanently disadvantaged when the French inventor Antoine Redier patented an adjustable mechanical alarm clock in 1847.

Not all technological unemployment has been quite so obscure as the lonely death of the knocker-up. The economist Gregory Clark has convincingly argued that the working horse was one of the biggest victims of the invention of the internal combustion engine. According to Clark, there were 3.25 million working horses in England in 1901. By 1924, less than a quarter of a century later, that number had been reduced to fewer than 2 million: a steep drop of 38 percent. While there was still a use for horses plowing fields, pulling wagons and working in pits, the arrival of the internal combustion engine had driven down costs far enough that the wage for this work was so low it often wouldn't even pay for a horse's feed.

As machinery became more and more advanced, this trend picked up speed through the twentieth century and beyond. Thanks to dual advances in both Artificial Intelligence and its sibling field of robotics, automation is now sweeping across more industries than ever. In warehouses, robots are increasingly used to select and box up products for shipping. In the service industry, robots are used to prepare food—and even serve it to customers. To whit, the San Francisco startup Momentum Machines, Inc. has built a robot capable of preparing hamburgers. Current models can prepare around 360 per hour, and are capable of doing everything from grinding the meat for the burgers and toasting the buns to adding fresh ingredients such as tomatoes, onions and pickles. Another company, Infinium Robotics, constructs flying robot waiter drones, which navigate around restaurants using infrared sensors, and can carry the equivalent weight of two pints of beer, a couple of glasses of wine and a pizza.

The advantage of these machines is obvious. While research and development costs outstrip those of training a human, once this has been done they cost just a fraction of what a person would charge to carry out the same task. As the BBC assures us about waiter drones, they are "sturdy, reliable, and promise never to call in sick at the last minute." Alexandros Vardakostas, the cofounder of Momentum Machines, puts it even more bluntly: "Our device isn't meant to make employees more efficient. It's meant to completely obviate them."

The use of smart devices of the type described in chapter three is also having a significant impact on certain types of employment. In the US city of Cleveland, councils have distributed special bins, equipped with radio-frequency identification tags, to residents. Thanks to the technology, city crews are able to see whether residents are putting their garbage and recycling out for pickup. As a result, Cleveland eliminated ten pickup routes and slashed its operating costs by 13 percent. Although this is a net positive for efficiency, fewer pickups also means that fewer garbage collectors are needed.

The most unexpected shift when it comes to AI's impact on employment, though, is what it means for white-collar jobs that don't require manual labor. The tasks that today's machines are getting better at carrying out instead involve cognitive labor, in which it is our brains that are being replaced, not our bodies. This development was forecast by none other than Warren McCulloch—one of the inventors of the neural net—back in 1948. Speaking at an event called the Hixon Symposium on Cerebral Mechanisms in Behavior, at the California Institute of Technology, McCulloch told the assembled audience:

As the Industrial Revolution concludes in bigger and better bombs, an intellectual revolution opens with bigger and better robots. The former revolution replaced muscles by energy, and was limited by the law of the conservation of energy, or of mass-energy. The new revolution threatens us, the thinkers, with technological unemployment, for it will replace brains with machines limited by the law that entropy never decreases. These machines, whose evolution competition will compel us to foster, raise the appropriate question: "Why is the mind in the head?"

McCulloch's last point is the most pertinent one. The Industrial Age leaders of industry assumed it was their intelligence that would protect them from technological replacement. Manual work reduced men to flesh-and-muscle machines, thereby outdating them the moment superior machinery came along. But smart people? Industrial Age machinery wasn't likely to displace them any time soon, was it? Today's reality is somewhat different. As we've seen, the past few years have ushered in extraordinary advances concerning what machines are capable of. Machines have become not simply tools to increase the productivity of human workers, but the workers themselves. Computers are still at their best when it comes to dealing with routine tasks in which they follow explicit rules. However, advances in AI mean that the scope of what is considered routine has become far broader.

For example, a little more than ten years ago, driving a car was considered something that a machine would never do. This is because of the unstructured nature of the task, which requires the processing of a constant stream of visual, aural and tactile information from the immediate environment. That all changed on

October 9, 2010, when Google published a blog post revealing that it had developed "cars that can drive themselves." Kitted out with laser range-finders, sonar transmitters, radar, motion detectors, video cameras and GPS receivers—along with some cutting-edge AI software—the cars can negotiate the chaotic complexity of real-world roads. To date, Google's fleet of Googlemobiles have driven around 1 million miles without causing an accident. The one serious accident they've been involved with happened "while a person was manually driving the car."

What does this mean for taxi drivers and long-distance truckers? One possible future can be glimpsed by looking at air travel. Around half a century ago, the flight deck on an airliner had seats for five highly skilled and well-paid individuals. These included a pair of pilots, a navigator, a radio operator and a flight engineer. Today, just two of those—the pilots—remain. And they may not be around for long, either. "A pilotless airliner is going to come; it's just a matter of when," said Boeing executive James Albaugh in 2011. No doubt seeing that as a challenge, Google has already created its Project Wing initiative, which seeks to extend its work on driverless cars to driverless commercial airlines. "Let's take unmanned all the way," Project Wing's leader Dave Vos said during a panel discussion at the annual conference of the Association of Unmanned Vehicle Systems International. "That's a fantastic future to aim for."

It's tough to predict how many other industries will be disrupted thanks to Artificial Intelligence, although we can make an educated guess. In 2013, a study carried out by the Oxford Martin School concluded that 47 percent of jobs in the US are susceptible to automation within the next twenty years. The authors predicted that there would be two main "waves" of this AI takeover. "In the first wave, we find that most workers in transportation and logistics occupations, together with the bulk of office and administrative support workers, and labor in production occupations, are likely to be substituted by computer capital," they wrote. In the second wave, every task involving finger dexterity, feedback, observation and working in confined spaces will fall prey to AI.

What is likely to surprise people is just how broad some of these categories may prove to be. AI has already made inroads carrying out many of the information-based tasks that are traditionally the domain of high-cognition professionals like doctors or lawyers. Lawyers, for instance, are being squeezed by the arrival of tools like LegalZoom and Wevorce, which use algorithms to guide customers through everything from drawing up contracts to filing for divorce. This kind of automation will particularly affect younger workers, such as junior lawyers, who previously learned their jobs by carrying out routine tasks like "discovery"—referring to the task of gathering documents that will be used as evidence in a court hearing. Thanks to e-discovery firms, this work can be done by machines for a fraction of the cost of paying an army of junior lawyers to do it. As a result, it is likely that many law firms will stop hiring junior and trainee lawyers altogether.

Even high-level executives may have to watch their backs, though. In 2014, a venture capital firm in Hong Kong named Deep Knowledge Ventures announced that it had appointed an AI to its board of directors. Given the same level of influence as human board members, the role of the Artificial Intelligence was to weigh up financial and business decisions regarding investments in biotechnology and regenerative medicine. At least according to its creators, the AI's strength was its ability to automate the kind of due diligence and historical knowledge of trends that would be difficult for even a human to spot.

Whichever way you slice it, work as we know it is about to change.

The Positives of Techno-Replacement

In 1589, the British inventor William Lee invented a stocking frame knitting machine. According to legend, he did this because the woman he was wooing showed more interest in knitting than she did in him. (Which, of course, begs the question of what kind of an outcome a person hopes for when they try to woo a beloved by putting them out of business?) Looking to protect his invention, Lee traveled to London and, at considerable expense to himself, rented a building with the aim of showing his machine to Queen Elizabeth I. The Queen turned up to the demonstration but refused to grant Lee the patent he was requesting. The words she used to explain her decision have gone down in history: "Thou aimest high, Master Lee. Consider thou what the invention could do to my poor subjects. It would assuredly bring to them ruin by depriving them of employment, thus making them beggars."

At the time, England had powerful guilds, which eventually had the effect of driving William Lee from the country. As the historian Hermann Kellenbenz has observed, these guilds "defended the interests of their members against outsiders, and these included the inventors who, with their new equipment and techniques, threatened to disturb their members' economic status."

It is highly unlikely that any government, regulatory body or (especially) venture capitalist firm in the UK or United States would act in the same manner today. A person applying for a patent will be scrutinized on the originality of his or her idea, not on the long-term impact it is likely to have on society.

However, despite a lack of people willing to behave as Queen Elizabeth I did, it may be the case that the long-term implications of AI are not as bleak for employment as some would have you believe. Yes, Artificial Intelligence's risk to our livelihood is one of the most pressing issues we need to examine, but there are also plenty of reasons to be optimistic.

Let's start with what may sound a controversial idea: that there is a moral imperative for getting rid of certain types of work.

To give an example most people can surely agree on, there were more than 1,000 chimney sweeps employed in Victorian London. Unlike the romantic picture presented in movies like Disney's *Mary Poppins*, chimney sweeps endured a brutal existence. Children were often used as sweeps because their small frames allowed them to go down the narrowest chimney stacks, where adults could not reach. Child chimney sweeps started working as young as three years old. Since most would literally outgrow the job by nine or ten, some bosses underfed their employees so that they could continue fitting down chimneys. Death could result from chimney sweeps falling down chimney stacks, or getting stuck in them without anyone knowing what had happened—leading to death from exposure, smoke inhalation or even burning. Many children suffered irreversible lung damage from constantly breathing in soot.

Regardless of our views on youth unemployment, not too many would be in favor of bringing back child chimney sweeps today. Technology has replaced our need for people to perform this role, both through smarter power-sweeping brushes and, more important, through the replacement of coal and wood burners by gas and electric heating as our primary means of staying warm. Despite putting a number of people out of work, it is fair to say that this type of technological unemployment was a positive one. This isn't an entirely new realization. Writing in 1891, at the very height of the Victorian era, Oscar Wilde argued: "All unintellectual labor, all monotonous, dull labor, all labor that deals with dreadful things, and involves unpleasant conditions, must be done by machinery. On mechanical slavery, on the slavery of the machine, the future of the world depends."

Today, an equivalent type of "monotonous . . . labor" that takes place in "unpleasant conditions" might be the manufacturing of devices like our smartphones and tablets, which are regularly carried out in places like China and India. The pristine white boxes our iPhones arrive in, complete with the sunny slogan "Designed by Apple in California," make it easy to forget that what we are holding are, in essence, Industrial Age products pieced together in Eastern factories under sometimes tough and unpleasant conditions.

One of Apple's largest manufacturers is a Taiwanese company called Foxconn. Foxconn operates on a scale that is unimaginable to many of us in the West. As the single largest private sector employer in China, it employs around 1.4 million people: roughly equal to the total population of Glasgow. Foxconn's factories are more like giant campuses than they are factories as we might think of them. The factory workers live and work there, sleeping in multi-person dormitories before trudging to work to spend hours on a conveyor belt line. Foxconn has frequently come under fire for its treatment of human workers. In 2012, living conditions at a company factory in Taiyuan, in China's northern Shanxi province, were reportedly so poor that they sparked a riot. There have also been multiple instances of suicides among Foxconn workers, which has led to suicide nets being erected outside the factories and dorms.

If we could replace this work with automation, should we be morally obligated to do so? Perhaps so. And perhaps we will. In 2011, Foxconn's CEO Terry Gou announced plans to replace 1 million of Foxconn's factory workers with manufacturing robots, known as "Foxbots." Like many of the predictions we've seen about the speed at which such breakthroughs are possible, Gou's initial estimates were off. Having suggested that the robot replacements would be complete by the close of 2014, at the end of that year Foxconn was still hiring humans while reporting problems with the production accuracy of its Foxbots. Terry Gou has now revised his estimates to 2016 until the Foxbot army is ready to take over manufacturing on devices like the iPhone. Although Foxconn is likely not developing Foxbots for ethical reasons so much as financial ones, the net result is still an ethical one in terms of rendering unpleasant jobs obsolete, even if it then opens up the problem of what to do with the newly unemployed workers (see page 139).

There are other illustrations, too, referring to areas we may not even currently view as fraught with ethical challenges. At present, an average of 43,000 people die in the United States each year due to traffic collisions. That's a higher figure than those killed by firearms (31,940), sexually transmitted diseases (20,000), drug abuse (17,000) and other leading causes of death. Advances in AI and automation will certainly help to cut down on these deaths. Tesla chief executive Elon Musk has argued that, once we reach the point where self-driving cars are widespread, it would be unethical to continue letting humans drive vehicles. "It's too dangerous. You can't have a person driving a two-ton death machine," he said during an appearance at an annual developers conference for Nvidia, a Silicon Valley company which specializes in computer vision. Musk thinks the transition will take some time due to the number of cars already on the road, but feels that it could happen within the next two decades. The toll on taxi or truck drivers might be a negative in the short term, but getting as many human drivers as possible off the roads may turn out to be a positive move in the end.

Out with the Old, in with the New

Of course, ethical concerns don't mean anything if the choice is between doing a dangerous or unpleasant job and not being able to feed yourself and your family. It wouldn't be enough to ban child chimney sweeps in Victorian England if the government wasn't also going to provide children with free education and a chance to better their employment opportunities. Getting rid of undesirable jobs is only good if we can replace them with something else. Fortunately, AI can be of service here, too. Although it is certainly true that technological advances have classically displaced certain types of work, they have also created them.

For instance, the invention of the horse-replacing internal combustion engine sparked a shift that transformed countries like the US from agrarian economies—based on the farming of crops and cattle—to industrial ones. Two centuries ago, 70 percent of American workers lived on farms. Today, automation has eliminated all except 1 percent of these jobs, with machines taking the rest of the work. Those workers didn't become members of the long-term unemployed, though. Instead they moved to rapidly expanding cities and got jobs in factories.

This is what economists call the "capitalization effect," in which companies enter areas of industry in which both demand and productivity are high. The result is new quantities of previously unimaginable employment, able to offset the destructive effects of economic shifts. There is no compelling reason to believe that we won't see a similar transition in the modern AI age. As with the shift from agrarian to industrial economy, we will witness a similar number of current jobs disappear within most of our lifetimes. However, digital technologies will also create a variety of new job categories, many of which were unimaginable just a few decades ago.

Consider the meteoric rise of content generators who make a living thanks to YouTube. In 2014, the popular YouTube star Felix Kjellberg—better known by his online moniker PewDiePie—earned $7 million from gaming commentary videos. With a subscriber count in excess of 37 million, PewDiePie is the star player in a growing "vlogger" job category, which didn't exist until 2000, and only took off in 2005.

PewDiePie's success is exceptional, but it's part of a bigger story. More than 1,500 new types of occupation have all appeared as official job categories since 1990. These include roles such as software engineers, search engine optimization experts and database administrators. The use of AI within video games has meanwhile inspired millions of fans to seek out work as professional game developers. Like "vloggers," the job of video game designer was not the dream of a single person 200 years ago, although today the video game industry is among the world's most valuable entertainment industries. The launch of *Grand Theft Auto V* in September 2013 achieved worldwide sales of more than $634 million—becoming the biggest launch of any entertainment product in history. By 2017, it is estimated that the video game industry will be valued at $82 billion globally.

In 2014, 6 percent of the UK workforce was employed in one of these new job categories. This concentration is at its highest in major cities. In central London, such roles accounted for 8.6 percent of all jobs in 2004, increasing to 9.8 percent a decade later. As with many new consumer technologies, there is evidence that these new job categories start out with early adopters and entrepreneurs in cities, before diffusing to other regions as they become established.

The Revenge of the Mechanical Turk

These jobs don't just involve building bigger and better AI systems, but also working alongside them. The latter roles are sometimes called Mechanical Turk jobs, named after a chess-playing automaton called "The Turk," built in the eighteenth century by the inventor Wolfgang von Kempelen. The Turk toured Europe, where it beat talented chess players, including Napoleon Bonaparte and Benjamin Franklin. However, it was later revealed that the Turk was not really a machine at all, but rather a human chess master controlling the operations of a puppet-like "robot." Much the same is true of today's AI tools, which appear to be examples of 100 percent machine intelligence but are, in fact, a sort of hybrid intelligence requiring the input of both humans and machines at every stage. A Mechanical Turk job applies to any jobs that are assigned to humans because machines are not yet capable of carrying them out. As a result, this is sometimes described as "Artificial Artificial Intelligence."

Many companies have experimented with AAI. The best known of these is Amazon's MTurk platform, which allows individuals and businesses to crowdsource humans to carry out what are known as HITs, standing for Human Intelligence Tasks. This could be anything from labeling the objects found in an image to make searching easier to transcribing audio.

Amazon is far from alone. In August 2015, Facebook launched "M," a text-based AI assistant, similar to the technologies described in chapter four. Unlike Siri, Google Now and Microsoft's Cortana, M uses a combination of both human and Artificial Intelligence to answer user queries. If the AI is unable to respond to a question satisfactorily, humans can take over the conversation.

Twitter also employs a large number of contract employees, called judges, whose job it is to interpret the meaning of different search terms that trend on the microblogging service. For instance, at 6:00 p.m. on October 3, 2012, Twitter experienced a sudden spike in US searches for the phrase "Big Bird." Using its human judges, Twitter was able to determine that this was a reference to Mitt Romney (who was talking about government funding for public broadcasting) and not an explicit search for *Sesame Street*. Why were humans better than machines for this job? Because we understand oblique references more easily than machines do.

As Twitter engineers explained in a blog post: "After a response from a judge is received, we push the information to our backend systems, so that the next time a user searches for a query, our machine learning models will make use of the additional information."

The need for these Mechanical Turk roles will only increase as companies invest in bigger and better AI systems. Amazon's MTurk system, for example, was described in 2011 as having an active user base of "more than 500,000 workers from 190 countries." It is likely that this number is significantly higher today.

The main criticism of Mechanical Turk systems is that, in many cases, the work is compensated very poorly. Even those Mechanical Turkers who live in the US currently make only around $1.60 per hour, with no worker protection or benefits. This is because the Human Intelligence Tasks—despite being hard enough to baffle many machines—are generally unskilled by human standards and therefore they are jobs that the majority of people are more than capable of performing. Because of this, the potential global supply of workers available to do the work is high, which drives down the cost. The net result is what at least one critic has labeled a "Digital Sweatshop."

That is, of course, if workers making today's AI systems smarter get paid at all. As we've seen so far, many of today's most successful AI applications rely on crunching millions or even billions of pieces of data generated by humans. The unwritten user agreement is that companies give their products away for "free" on the condition that they then get to use the resultant data to sell ads or make their AI systems smarter. For example, as with all of the above illustrations, Google's online translation service appears to be 100 percent machine intelligence. In reality it works based on data provided by human users, taking individual words and phrases that have been matched up previously in human translations and applying this knowledge to entire bodies of text. Next time you use Google Translate, consider for a second that some of the Mechanical Turks who make it possible are highly skilled human translators, often with PhDs in various languages.

Unlike the people who voluntarily sign up for MTurk tasks, these translators will never get paid anything for their contribution—other than the sum they were paid for carrying out the original contracted work, that is. Hanna Lützen may get paid by the Gyldendal publishing house for translating the *Harry Potter* books into Danish, but Google pays her nothing if those combined 1 million-plus words then help its system translate a love letter from your girlfriend in Denmark.

This differs from the legal issues surrounding similar "sampling" in the real world. For instance, in the world of hip-hop, music artists regularly chop up and reuse samples of songs by other musicians. When they do this, they have to pay for these samples to be cleared. If they fail to do so, legal action can follow, as it did in 2006 when a judge ordered that sales of the Notorious B.I.G.'s album *Ready to Die* be stopped because it used an excerpt from a 1972 song, entitled "Singing in the Morning," without the proper permission. The German electronica band Kraftwerk has successfully argued in court that even the smallest samples of sounds—such as a few bars of a drum beat—are protected by copyright.

What does this have to do with the legal use of data? All of us are now Mechanical Turkers to some extent, since the data we help generate on a daily basis is what makes AI systems smarter. Whether it's uploading photos to Facebook or typing in a block of twisted letters to prove our humanity to a CAPTCHA, we're all helping to train the robot successors who are after our jobs. At some point in the near future, a serious conversation needs to be had about the value we place on data. If, as is often said, data is the oil of the digital economy, then we need to place a proper valuation on it.

Virtual reality pioneer Jaron Lanier has suggested one way to do this would be a universal micropayment system. Lanier has given a few illustrations of how this might work. Imagine, he suggests, that you sign up for an online dating service where the data you provide to refine your own romantic matches also helps the company perfect its algorithms for attracting other users. Or if Facebook uses your profile picture in an ad to target a page to one of your friends. Another example might be Netflix using your viewing preferences to help commission a show like its Emmy award–winning *House of Cards*, which was created entirely on the basis of Netflix user data.* In cases like this, a formula could be established to determine both where data originated and how important the data was in shaping certain decisions. This calculation would then necessitate a micropayment being made to users in the same way that a royalty rate is paid to a musician whose work is sampled by another artist.

The idea sounds far-fetched, but the law is still catching up on many of the technological shifts we've seen in the past decade. Precedents like the European Union's "right to be forgotten" ruling against Google show how laws are still catching up with the realities of new digital technology. At some point, the question of data ownership is sure to come under scrutiny. To return to the music sampling analogy, a large number of cases of illegal sampling went under the radar in the early days. It was only later on, when the technique became part of mainstream music, that artists suddenly found themselves in court facing multimillion-dollar fines for copyright infringement. Similarly, as the AI-driven shift in employment makes job categories like Mechanical Turkers

---

*For more on Netflix's approach to data-driven algorithmic creativity, check out my previous book *The Formula: How Algorithms Solve All Our Problems . . . And Create New Ones*.

more prevalent, conversations need to be had about who owns the data driving AI systems. Implemented correctly, there's no reason this shouldn't aid companies as well as individuals. The real value in many twenty-first-century businesses is the analyzable data they hold. If users were financially compensated for feeding data into these businesses, it would add an extra incentive for using them. If the kind of universal micropayments Jaron Lanier describes were applied to every piece of data we generate, it is not unthinkable that Mechanical Turkers could go from making $1.60 per hour to earning an amount closer to the UK minimum wage of $10.72, or even more. This would be a key step in establishing a digital framework in which AI systems get smarter, but humans are able to share in the wealth created.

The Human Element

Mechanical Turk jobs involve humans working behind the scenes in AAI roles which are often hidden from view. However, as AI becomes a larger part of all our lives, a number of companies have started emphasizing—rather than downplaying—the role humans have to play in their systems. Like Google, Facebook and other tech companies, Apple has competed fiercely to hire AI experts in recent years. According to a former Apple employee, the company's number of machine learning experts has tripled or quadrupled in the past several years. As with these other companies, Apple uses humans as part of its largely AI-driven services. However, unlike the other companies, Apple presents its human workers as a selling point, not simply as a stand-in for the bits of its technology that don't quite work properly yet. When Apple introduced its much-anticipated Apple Music streaming service in June 2015, one of its most heavily advertised features was its reliance on humans with specialist music knowledge to curate playlists. "Algorithms are really great, of course, but they need a bit of a human touch in them, helping form the right sequence," executive Jimmy Iovine told the *Guardian* newspaper shortly after Apple Music's launch. "You have to humanize it a bit, because it's a real art to telling you what song comes next. Algorithms can't do it alone. They're very handy, and you can't do something of this scale without 'em, but you need a strong human element."

In reality, algorithms *can* sequence music in a way that is palatable to many users. AI tools are able to generate playlists based on genre, era, artist, tempo, or countless other metrics. A number of companies (Apple included) have even explored technology such as mood-detecting headphones, so that music tracks can be selected for users based on whether they happen to be out jogging or lazing on the sofa. But what Apple astutely noticed was that humans enjoy interacting with other humans. Apple Music's human curators are not invisible parts of the algorithm process, but flesh-and-blood experts with the goal of helping you discover music an algorithm would have been unlikely to recommend. An AI can only recommend music to you based on the data it has about your previous favorite songs or what is popular with other listeners. A human expert, on the other hand, can do more than that.

Apple's expert "tastemakers" include names like popular DJ Zane Lowe, who left his high-paying job at BBC Radio 1 for a starring role in Apple's new streaming music service. Others include former NWA rapper Dr. Dre and pop star Elton John—none of whom are low profile, or (presumably) working for $1.60 per hour. Depending on how big Apple Music gets, it is astonishing to think that a tech company could wind up being one of the big employers of human DJs on the planet.

This new focus on human traits like creativity and social intelligence will only become more important as AI gets smarter. Although Artificial Intelligence is becoming better at communicating in a humanlike way and is proving surprisingly creative in certain applications (as we shall see in the next chapter), these are skills that will remain prized in humans.

Observing this transition, Harvard University's economics professor Lawrence Katz has coined the term "artisan economy." Artisans are skilled workers who often carry out their work by hand. During the Industrial Revolution, artisans were increasingly replaced as automation took over. For example, mechanical looms took jobs away from artisans skilled at handcrafted artisanal weaving. Today, there is evidence that trend is being reversed.

When Katz talks about an artisan economy, he doesn't just mean weaving, of course. "Artisan economy" means the return of products that are not machine-driven and homogenous, but rather rely on human creativity and interaction. For instance, the carpenter who sells and fits standardized products will struggle with the rise of technologies like 3-D printing. However, the carpenter who is able to assess their customer—working out what it is that they're going to want to use a new cabinet or desk for and adapting their work to suit—will fare much better. Similarly, care workers who are emotionally checked-out and little more than babysitters could conceivably be replaced by robots. An artisanal dementia coach or home health aide—full of bright ideas to keep clients engaged—has the potential to flourish, particularly in a market with a growing elderly population. Much the same can be said for inspirational human personal trainers going up against smart wearable devices, human taxi drivers with insider knowledge of good places to visit going up against self-driving cars, and empathetic lawyers going up against services like Wevorce.

These artisan economy jobs are likely to be overwhelmingly "high-touch," meaning that they rely on personal contact. This makes it tougher for them to be replaced by outsourcing, robots, or the right algorithm. Unlike the artisans of the Industrial Revolution, though, today's workers in the artisan economy can use technology to augment, rather than replace, their employment opportunities. Scaling a business to reach millions, or even billions of people, is possible in a way that it never was before the digital age. In 2014, a story appeared on *Business Insider* about an SAT tutor who charges $1,500 for ninety minutes of one-on-one tutoring—carried out via Skype. Even in an age of educational apps and online learning tools, the tutor was able to command incredibly high prices due to his proven ability to raise test results.

Another example of the artisan economy at work is Etsy, the online marketplace where people can sell handmade or vintage products. Having launched in 2005, Etsy currently offers more than 29 million different pieces of handmade jewelry, pottery, clothing and assorted other *objets d'art*. By 2014, gross merchandise sales for the site had reached $1.93 billion, with sellers taking home the vast majority of this. Trading on the popularity of artisanal goods, some sellers have proven incredibly successful, earning many thousands of dollars each month. Despite the site's success, the focus on handmade artisan goods remains central. When it was revealed that one Etsy store owner was bringing in upward of $70,000 a month selling headbands and leg warmers that turned out to be mass-produced in China, there was an immediate uproar from the community.

Working out which tasks make lasting business sense in the artisan economy will be a matter of trial and error. It will likely be areas where non-machined irregularities are valued, such as the personal trainer who offers a personalized service to clients, or the executive who does more than just crunch numbers. In other fields, we will prove less willing to hand back tasks previously given to machines, however. No one wants humans instead of machines to build their cars anymore. Irregularities are a lot less appreciated when they happen on the motorway at high speeds.

To cope with this paradigm shift we will also need to do better at training the new generation. Currently, education is stuck in the same Industrial Revolution paradigm it has been in for more than 100 years. In an age in thrall to the factory, it followed that schooling borrowed the same basic conveyor belt metaphor that was then being used to churn out identical Model T Fords. Standardized lesson plans were designed to teach students specific skills for pre-prescribed roles in the workplace. This standardization assumed that the skills students were learning were unchanging ones that they would rely on for the rest of their lives. In today's world, learned skills routinely become obsolete within the decade they're learned—meaning that continual learning and assessment is needed throughout people's lives. In an age in which we have the Internet on every smartphone, we will also need to question the purpose of teaching children to mentally store large amounts of information through uninspired rote learning.

Barring some catastrophic risk, AI will represent an overall net positive for humanity when it comes to employment. Economies will run more smoothly, robots and AIs will take over many of the less desirable jobs and make new ones possible, while humans are freed up to pursue other, more important goals. Artificial Intelligence may be able to do a lot of the jobs we currently do—but humans are far from irrelevant.

After all, several years after Ken Jennings was roundly beaten by IBM's Watson AI, we're not yet letting our dinners grow cold to go and watch two AIs battle it out in trivia shows on TV. Despite the braininess on show in an episode of *Jeopardy!*, it's the human personalities the audience really wants to see. This drama is ultimately what matters the most.

6

Can AI Be Creative?

IN THE LAST chapter, we saw how creativity is likely to remain one of the areas where humans will have the edge over machines when it comes to employment. As Artificial Intelligence continues to "eat the world" (as venture capitalist and software engineer Marc Andreessen believes), it is those jobs involving human creativity that are likely to withstand the march of automation.

That's not to say AI doesn't have the capacity to be creative, however. In June 2015, Google unveiled its Deep Dream project. A fascinating research project from a company that typically cares far more about engineering than aesthetics, Deep Dream is an AI-driven image generation program that works by tapping into the well of pictures Google has indexed over the past decade and a half. At its digital fingertips, Google has what is almost certainly the largest archive of images ever assembled under one umbrella. In 2001, the company had 250 million images indexed and made searchable to its users. In 2005, that number had grown to 1 billion. By 2010, the index had jumped again, this time to 10 billion images. Today it is significantly more.

As we saw in chapter two, Google's use of deep learning neural nets has let its machines intelligently recognize the contents of individual pictures. To recognize what a chair is, for instance, Google's programmers show its neural network millions of pictures of chairs. After several million images, the neural network has established what is and isn't a chair. It's seen so many chairs that, one imagines, if it really wanted, it could draw chairs in its sleep.

Which, in this case, is exactly what Google planned.

Usually, Google uses its image recognition neural networks to classify images so that this does not have to be done manually. For instance, Google Photos allows users to type in a search term like "skyscrapers" or "graduation," which then immediately sets its neural network off looking for tall blocky buildings or mortarboards. With Deep Dream, the team imagined that the processes it typically uses to sort and recognize images could also be used to generate them from scratch. The idea was that, having looked at hundreds of thousands of different chairs from every conceivable angle, Google's neural network should not only be able to recognize a chair, but to reproduce the perfect Platonic form of a chair: what Ben Stiller's Derek Zoolander character might describe as the "essence of chairness." Rather than being based on one particular chair it has seen, Google is distilling everything it knows about chairs into a single new creation.

Or that was the idea at least. In the case of Deep Dream, Google's engineers were actually taking advantage of a fun quirk of its image-generating neural network. Left unassisted, the neural network had become confused: it discovered unusual relations in Google's 10 billion–plus images that make it difficult to work out where an object starts and stops. As a result, Google's Platonic objects (that perfect "essence of chair") sprouted some unusual appendages, such as long fleshy arms that hung from Deep Dream's idealized dumbbells like pink lengths of rubber tubing. As Google software engineers Alexander Mordvintsev and Mike Tyka pointed out in a blog post: "There are dumbbells in there all right, but it seems no picture of a dumbbell is complete without a muscular weightlifter there to lift them. In this case, the network failed to completely distill the essence of a dumbbell. Maybe it's never been shown a dumbbell without an arm holding it."

Normally, Google would correct what it called "these kinds of training mishaps." With Deep Dream it decided to go in the opposite direction. The result was surrealistic landscapes which seemed to owe more to Salvador Dalí or H. P. Lovecraft than Google cofounders Larry Page and Sergey Brin. The team allowed the neural network to accentuate whatever eccentricities it discovered. Instructed to maximize the elements found in each image, Deep Dream created trippy flights of fancy. Given an image and asked to classify it and then add more detail, the neural network became trapped in strange, fascinating feedback loops. Clouds were associated with birds, and Deep Dream sought to make them ever more "birdlike." A photograph of a clear sky would rapidly be filled with Google's idealized avians, as though the world's most powerful search engine had suddenly decided to become a graffiti artist. The same happened with whatever photos you showed it, as Google's neural networks created entire fantasy worlds out of nothingness. Trees became ornate buildings, leaves became insects, and empty oceans became alien cityscapes.

Deep Dream may not receive a Turner Prize nomination this year, but make no mistake about it: this was creativity, Google style.

A Robot Symphony

There is a wonderful scene in the 2004 movie *I, Robot*, in which the protagonist, played by Will Smith, has a conversation about computational creativity. "Can a robot write a symphony?" he asks. "Can a robot turn a canvas into a beautiful masterpiece?" The robot with whom he is conversing deflects the question by asking, "Can you?"

At first glance, asking if a machine can be creative isn't a matter of particular importance. Compared to some of the other issues addressed in this book—the future of employment or artificially intelligent medicine—it's easy to think of it as downright minor. It's the kind of question that two software engineers at Google might casually chat about over beers on a Friday afternoon.

In fact, artificial creativity is one of the most important issues faced in AI. Questioning whether Google's Deep Dream project is art may not be of interest to everyone, but the other implications of creativity are enormously significant. "Not until a machine can write a sonnet or compose a concerto because of thoughts and emotions felt, and not by the chance fall of symbols, could we agree that machine equals brain," argued the respected British neurosurgeon Sir Geoffrey Jefferson in 1949. Were a machine able to do this, should we therefore consider it intelligent?

Few subjects cause more consternation than the idea that a machine might be considered creative. A typical argument made by detractors is that art is created by emotions as opposed to intellect. Artists are viewed as irrational, while computers are the epitome of hyperrationality. *Time*'s Lev Grossman opines, "Creating a work of art is one of those activities we reserve for humans and humans only. It's an act of self-expression; you're not supposed to do that if you don't have a self."

Consider the regular sci-fi trope of the AI that gains access to human emotions, but becomes less useful (or even downright dangerous) as a machine as a result. For this reason, the history of AI research shows relatively little focus on recreating emotions inside a machine until recently. Unchecked emotion leads to impulsiveness and irrationality, and the goal of AI is to build rational, reasoning machines. As we saw in chapter four, computers are now getting better at recognizing emotions in users, but we're still far away from machines having emotions themselves. And maybe that's all the better for it.

But is the same true of creativity? Is this still some far-off mission, or is it much closer than we might think? Could it even be here right now? Looking at some of the images generated by Google's Deep Dream project, I found myself wondering how I would react were I to see them on a visit to the Tate Modern in London. Would I instantly dismiss them, or would I, perhaps to my embarrassment, struggle to separate them from whichever artwork had won that year's highfalutin art prize? Would we react differently if we knew that it was a data-driven creation from Google's neural networks than we would if we thought it was the work of a respected (human) artist? What if two identical pictures had happened to be created: one by a person, the other by an AI? Would we judge these differently?

Depending on your subjective appreciation of its images, Deep Dream might answer Will Smith's question about whether an AI can turn a blank canvas into a "beautiful masterpiece." As for his query about whether or not a robot can write a symphony, in July 2012, the London Symphony Orchestra took to the stage to perform an algorithmically generated composition entitled *Transits— Into an Abyss*. Developed by researchers at the University of Malaga in Spain, the symphony was described as "artistic and delightful" by one reviewer.

These are far from the only illustrations of AIs being used in the creative process. Go and see virtually any Hollywood blockbuster and the CGI effects used to bring non-human characters to life rely on techniques developed by Artificial Intelligence researchers. Animators at studios like Pixar animate their scenes using algorithms for automatically generating the realistic motions of on-screen virtual characters. Today's human animators aren't there to animate every frame, but rather to behave like movie directors: describing the scene in broad strokes, then leaving many of the subtle performance decisions to the actor. This actor is often an AI algorithm computing sequences of behavior for the virtual characters in a film, allowing them to reach their user-designated goal position in a scene. If humans are considered creative when they act, then why shouldn't virtual actors be regarded in the same way? Both are told what to say, and where to begin and end a scene. The creative bit is every *other* decision.

With that being the case, it's surely no surprise that AIs are also becoming adept at telling their own original stories. Here, for example, is a story written by "Scheherazade," an AI created by researchers at the Georgia Institute of Technology, named after the storyteller in *Arabian Nights*:

> John took another deep breath as he wondered if this was really a good idea, and entered the bank.

John stepped into line behind the last person and waited his turn. When the person before John had finished, John slowly walked up to Sally.

The teller said, "Hello, my name is Sally, how can I help you?"

Sally got scared when John approached because he looked suspicious. John pulled out a handgun that was concealed in his jacket pocket. John wore a stern stare as he pointed the gun at Sally. Sally was very scared and screamed out of fear for her life.

In a rough, coarse voice, John demanded the money. John threw the empty bag onto the counter.

John watched as Sally loaded the bag and then grabbed it from her once she had filled it. Sally felt tears streaming down her face as she let out sorrowful sobs.

John strode quickly from the bank and got into his car, tossing the money bag on the seat beside him.

John slammed the truck door and, with tires screaming, he pulled out of the parking space and drove away.

Okay, so it's not exactly *War and Peace*, but it doesn't take enormous generosity to say that it's no worse than your average holiday thriller novel. And it's getting better and better all the time. The goal of Scheherazade is to generate stories by researching details about different topics and transforming these into narrative plot points, much as a human author would do. The data comes from crowdsourced descriptions of scenarios, which the AI then aggregates and turns into original content. "If we ask it about bank robberies, we don't tell it anything about banks or robberies," creator Mark Riedl explains. "We don't even tell it what actions can be performed in banks. It has to get this knowledge from somewhere, so it asks humans to describe examples in natural language. It learns from those examples and then creates something new."

Don't get too hung up on the style of Scheherazade's prose, either. Recent research from the team at Georgia Tech demonstrates how the system's use of language can be modified, making it more terse or expressive according to what is being asked for.

Being Original

All of these examples of computational creativity are, of course, very different. They rely on different AI techniques and relate to different creative disciplines. Before we go any further, it is therefore important that we define exactly what we mean when we talk about creativity. This is easier said than done. As with so much about the human brain, unraveling the roots of human creativity remains a mystery to us. The first attempt to talk about this subject was an 1891 book, *The Man of Genius*, by an Italian physician named Cesare Lombroso. Lombroso linked extreme creativity with both genius and insanity. Associated traits included left-handedness, stammering, celibacy, precocity and neurosis: only one of which could genuinely be applied to computers. Today, neuroscience has taken us somewhat closer to answering the question of where creativity comes from, but we are still far from finding a definitive answer. We can, however, at least go some way toward establishing a broadly agreeable definition for creativity as a quality.

The first distinction to make is that creativity is not simply the act of creating. If it was, there would be no argument about whether AI can be creative, since the answer would be an unequivocal "yes." At its most basic level, a computer algorithm is a means of turning inputs into outputs. By this definition, virtually every piece of computer software in existence is creative, in much the same way that some linguists argue that language itself is "creative."

We might therefore modify our thesis to suggest that creativity is the act of creating something new. This definition requires further clarification because it immediately prompts the obvious follow-up question: "new to whom?" Every person in the world is creative to some degree, although not all of this personal creativity is perceived as equally valuable by society. For instance, children regularly come up with ideas that are new to them, but are well known to everyone else. Parents reward this behavior because it shows that their children are learning, but the fact that little Tommy has learned how to open a door or draw a figure eight is unlikely to be of lasting interest to the general public because it is nothing they haven't seen before. As a parallel example, I could tell you that I had the idea for a smartphone with a touch-based interface and its own app store, but unless I can prove I had it before Apple, the company's lawyers aren't likely to lose too much sleep over it.

So let's amend our definition once more and say that creativity is the act of creating something that is new to society as a whole. It is this type of creativity (which we might also call novelty) that is recognized when we talk about "being creative." Skill is an important part of the equation, but it's not everything—which is why the street artist who paints a watercolor of Leicester Square probably won't find it hung in a museum, even though it might be a technically skilled reproduction.

The Beatles Bot

Lior Shamir believes that novelty is to creativity what perspiration is to inspiration. In other words, if you can somehow figure out how novel an idea is, you can use that knowledge to come up with creative ideas.

Shamir is a computer science professor at Lawrence Technological University near Detroit, Michigan. His experiments into whether an AI can be creative started, simply enough, with a 2014 project. Shamir wanted to see if he could write an algorithm capable of charting the evolution of a band's sound as it progressed over the years. Shamir chose one of his favorite groups, the Beatles, for the experiment. He started by populating a database with samples from the band's music, taken from each of the Beatles' thirteen albums. Letting the computer analyze the songs revealed 2,883 unique numerical content descriptors, noting everything from pitch and tempo to other patterns we don't commonly associate with music. Shamir then used a statistical tool called the K-Nearest Neighbor algorithm to determine the measure of similarity between any two songs in the database. Without human intervention, the algorithm managed to sort all thirteen albums into chronological order, beginning with 1963's *Please Please Me*, before proceeding to *With the Beatles, A Hard Day's Night, Beatles for Sale, Help!, Rubber Soul, Revolver, Sergeant Pepper's Lonely Hearts Club Band, Magical Mystery Tour, The White Album, Yellow Submarine, Let It Be* and—finally—*Abbey Road*.

To make sure he wasn't imagining things, Shamir tested his algorithm on the music of other popular groups, such as ABBA, U2 and Queen. In each case, his software was able to sort the albums into the order in which they were recorded, despite having access to no information other than the music itself.

This is undoubtedly impressive, but for Shamir it is just step one in the process. With the musical DNA of the Beatles broken down into 2,883 different descriptors, Shamir thinks that it will soon be possible to generate entirely new songs that sound as though they could have belonged on, say, the Beatles' *Revolver* album.

"The computer will be able to compose songs based on the heuristics that you would find on that album," he says. "That may mean using the same scales, time signatures, musical instruments, or whatever. Once you're able to determine what makes a song novel, then generating that same novelty is just a matter of computing cycles. You might not wind up with a hit immediately, but have the computer generate it again and again and again, and you'll eventually get there."

What Shamir is suggesting isn't completely new. In 1821, a Dutch inventor named Diedrich Winkel built a mechanical device called the Componium, which now belongs to the Brussels Museum of Instruments in Belgium. Winkel had already found success by coming up with a new weaving machine which could produce a near-infinite number of different fabric patterns. The Componium was an attempt to transfer that same approach of algorithmic composition to the world of music. His creation was an automatic organ consisting of two barrels which revolved simultaneously. These barrels took it in turns to perform two measures of randomly chosen music while the other silently chose the next variation. A roulette-style flywheel acted as the "programmer" to decide whether or not a particular variation was selected. Winkel worked out that, if each performance lasted an average of five minutes, the Componium could play for more than 138 trillion years for every possible combination of music to be played.

Shamir's concept is different, since it wouldn't mean matching up different elements of existing Beatles songs, but rather generating entire new tracks in the style of the mop-topped group. Due to the computing cycles needed to generate this music effectively, he thinks this technology is currently around a decade away.

Could we be enjoying a new Beatles album by 2026, then? Shamir sees no reason why we shouldn't, at least not on a technical level. After all, at the time of writing, an advertisement on UK television depicts the late actress Audrey Hepburn promoting Galaxy chocolate bars. The ad shows a young Hepburn (or Audrey HepburnTM as she is credited) stuck on a bus in traffic on the Amalfi Coast in the 1950s. In her handbag is a bar of Galaxy, which wouldn't even start out as a brand until the following decade. Through the window, HepburnTM spots an attractive man in an open-top car, and quickly switches vehicles before tucking into the chocolate bar as the pair speed away. Rather than using existing footage and doubles, the ad entirely re-created Hepburn's face in close-up with the latest computer graphics. If such technology is acceptable to the general public in an ad now, will we one day feel the same way about the musical equivalent?

Using holographic technology, it would even be possible to

have an era-appropriate version of the Beatles perform their new material on stage. This is exactly what happened when Michael Jackson was reanimated for an appearance at the 2014 Billboard Music Awards, to perform his posthumous hit "Slave to the Rhythm," along with an army of flesh-and-blood backup dancers. Imagine the equivalent with John, Paul, George and Ringo reuniting to sing a song that they didn't write—but could have— in 1967. Using Lior Shamir's algorithm to show how the Beatles' musical style changed during the decade they were around, perhaps we will be able to predict the sort of songs the Beatles might have written had they stayed together until 1975, 1983, or even today.

"I think it's very obvious that computers are going to be creative," Shamir says. "There's no doubt about it. People like to mythologize creativity like it's just about a momentary flash of divine inspiration. It took Paul McCartney two years to finish the song "Yesterday." There's a process. Creativity is about heuristics. It's about evaluating different paths and decisions until the right one is discovered. That's where AI comes in."

A Flash of Genius

Aristotle dreamed of a machine that could be creative. In *Politics*, written around 350 BC, he described his wish for an instrument which could play itself, because this would mean the end of slavery, since slaves were needed to carry out the task of providing music. To Aristotle, the Componium probably would have been considered machine intelligence, since it was able to compose new tunes and play them unaided. But today we would not consider the Componium intelligent. For starters, its music was not truly composed from scratch, but rather limited to new combinations of the music provided in its barrels. Like the "expert systems" described in chapter one, whatever intelligence the Componium had (in this case, its knowledge of music) came from its human "programmers." Lior Shamir's AI composer is the same, although it is far more complicated than the Componium. It may not be performing variations on existing melodies, but the data on which it is based is the result of human creativity, much like Google's translation tools.

Imagine, if you will, that Lior Shamir was able to get his AI to generate a new album in the style of the Beatles, composed of entirely new songs but using the same high-level descriptors that make a Beatles song sound the way that it does. Who should get songwriting credit: the AI or Paul McCartney? Questions like this lurk in the near future. Similarly, it seems inevitable that at some point in our lifetimes a CGI character will win an acting award. If Audrey HepburnTM was reanimated to play the lead in an Oscar-winning summer blockbuster, do we give credit to the software, the animator, or the real Audrey Hepburn for providing the original data? These scenarios remain thought experiments for now, but they highlight an important point about computational creativity. Simply put, if humans are involved in the process, we will ascribe credit to them instead of the machine—just as we'll give Jackson Pollock credit for his influential drip paintings, rather than giving the nod to paint and gravity. When Kasparov lost to IBM's chess-playing computer Deep Blue, he commented that he saw deep intelligence and creativity in the machine's moves—hinting not that AI had developed these qualities, but rather that

IBM was somehow cheating by using human chess players behind the scenes.

Because most of today's AI systems rely on crunching huge amounts of previously existing data, it is possible to argue that all of the previous examples of computational creativity in this chapter rely on turning old data sets into new inputs. Even if the result is a new idea, the new idea is an iteration of older ideas, gathered from the data it has access to. This much is certainly true. However, it is a mistake to say that this is fundamentally different from the way that humans create. As humans, we like to mythologize human creativity, viewing great ideas as the result of flashes of genius that come out of nowhere. For instance, the American electrical engineer Harold Black came up with the idea for the negative feedback amplifier while taking a ferry ride to work across the Hudson River in New York City. The flash of genius was so profound a moment in his life that Black decided to call his autobiography *Before the Ferry Docked*, although he died before he could finish it. The Hungarian-American physicist Leó Szilárd came up with the concept of nuclear chain reactions while waiting for the red light to change at a pedestrian crossing outside London's British Museum. And so on. Journalists like myself add to this idea of invention being the result of flashes of genius. Finding out where a person was when they invented the computer mouse or wrote the novel that changed their lives makes the process feel both transcendent and, somehow, achievable, as though inspiration might strike any one of us at any moment.

But the reality is that creativity is never created in a void, and is always about new ways of combining old ideas. Writing in the 1600s, the influential English philosopher Thomas Hobbes described imagination as "decaying sense." In other words, Hobbes believed that imagination rearranges our past sense impressions and the knowledge that is built on them. This is the same reason artists throughout history have subjected themselves to different emotional states, since this allows them to draw on previous experiences in the same way that a chef does with ingredients. If Hobbes' theory of imagination is correct, it would not substantially differ from the imagination of a machine.

To return to Lior Shamir's favorite band, the work of the Beatles was undoubtedly original, but it was inspired by the earlier work of artists like Chuck Berry, Carl Perkins and Elvis. As John Lennon said, "Nothing really affected me until I heard Elvis. If there hadn't been Elvis, there would not have been the Beatles."

Ideas that go completely against our expectations would prove bewildering. This is illustrated in a great scene in the movie *Back to the Future*, in which Marty McFly finds himself on stage at a school dance, having time-traveled back from 1985 to 1955. Armed with a guitar and a memorized catalogue of what, to a 1980s teenager, are classic songs, Marty "invents" Chuck Berry's 1958 song "Johnny B. Goode" by playing it several years before it was recorded. Euphoric at the crowd's reaction, he drifts into 1980s-style heavy metal, producing feedback by playing his guitar next to the amp. The kids stop dancing and look confused by what sounds to them like nothing more than noise. Marty comes to his senses. "Uh, sorry, you guys aren't ready for that," he apologizes. "But your kids are gonna love it." The sounds of classic rock and roll—which extrapolated ideas from the rhythm and blues music of the 1940s—make sense to a 1955 audience. Heavy metal music, which did the same to ideas from 1960s and 70s rock music, made no sense.

However, the act of simply combining ideas by themselves is

not necessarily creative, even if the result is novel. As Apple's CEO Tim Cook has said about unnecessary invention, "You can converge a toaster and a refrigerator, but you know those things are not going to probably be pleasing to the user." One of my favorite movies of all time is a film by Robert Altman called *The Player*, a satire on Hollywood and its sometimes lack of creativity. Throughout *The Player*, a running joke is the lazy shorthand descriptions that Hollywood insiders use to describe the different projects they are working on, which are always billed as "Movie *A* meets Movie *B*," with each referring to an existing popular hit. The joke is that the titles being mashed together to form new projects are totally diametrically opposed to one another. "It's *Out of Africa* meets *Pretty Woman*," says a screenwriter, pitching her script to a studio executive at the start of the film. Later, someone describes a "psychic, political, thriller comedy with a heart" that is "not unlike *Ghost* meets *Manchurian Candidate*."

Creating a computer program that could do the same thing—only much quicker—is well within our grasp. Using the list of roughly 328,952 feature films that appear on the Internet Movie Database, I could write a program that matches up every combination of films ever made. Within a short time of running it I would have more potential hits on my hands than a roomful of highly paid Hollywood screenwriters could come up with in a lifetime. Fancy a comedy about medieval Swedish knights competing to lose their virginity in a chess game with Death (*The Seventh Seal* meets *American Pie*)? Just send over the check!

As Lior Shamir points out, a computer is far better equipped than most of us to judge novelty due to the amount of data it has access to. For example, Google Books has scanned and digitized more than 30 million books, the contents of which it is now able to search. Google has estimated that there are approximately 130 million distinct books in the world, and has made clear its intention to scan all of these by the year 2020. Compare that number to the 25,000 books read by the woman who has laid claim to the title of Britain's most avid reader, having read around a dozen books each week since 1946. In an entire lifetime, even the most prolific reader is unable to read one-thousandth of the books Google has absorbed since it started its book-scanning project in just October 2004. With increasingly large data sets, computers are getting better and better at performing tasks like textual analysis, which is why they are being used for tasks like identifying who wrote particular books in cases where this is unknown.

But generating novelty is not enough. My movie title-generating bot would be prolific, but it would simply reverse the problem a lot of screenwriters have. Instead of not having enough ideas to choose from, suddenly we have far too many. It's still a data problem, just an inverse one. What makes someone creative is the ability to recognize that they are on the right lines with a certain idea. Shortly after he returned to Apple in 1997, Steve Jobs described innovation as the ability to say no to 1,000 possible ideas. "You have to pick carefully," he said. "I'm actually as proud of the things we *haven't* done as the things I have done." Steve Jobs eventually led Apple to create iTunes, the iPod, the iPhone and the iPad, but before he did this he said no to dozens of other products the company had been working on in his absence.

Fortunately, machines are getting better at this task, too. So long as we're able to tell them what we're looking for, they are able to create new imaginative solutions—even if means outperforming humans to do so.

Can an AI Be an Inventor?

Jason Lohn was thirty-one years old when he joined NASA in 1996. A trained electrical engineer, who has since gone on to work at Google, Lohn was given the job of designing antennae for use in spacecraft missions. "Antennae are extremely important in space," Lohn says. "If you don't have a good antenna system, you're basically launching a very expensive ball of metal, because we have no way of talking to it." The challenge with antenna optimization is how best to create a communication channel that is as high bandwidth as possible, while also being sufficiently small. Early antennae were capable of sending just a few bits (binary digits) at a time. One day the dream is to have full motion video streaming from space. Lohn was fully aware of the complexity of the problem, and he had an idea for solving it: why not hand over the design process to an AI?

"People had been using AI for tasks like scheduling and planning for years, but I wanted to use AI to improve the actual hardware that was being used in space missions," he says.

When he was an undergraduate, Lohn read Richard Dawkins' book *The Selfish Gene*, one of the most important books about gene-based evolution. "I was fascinated by the power of natural selection," he continues. In graduate school, Lohn began exploring the idea of replicating this kind of evolutionary process to solve design problems.

This is not, in itself, a new idea. For hundreds of years, mankind has steered evolutionary processes to breed new kinds of animals and plants that best suit our needs and wishes. During the eleventh century, the brilliant Persian polymath Abu Rayhan Biruni observed how foresters could create better trees by leaving branches they perceived to be excellent, while cutting away the rest. This notion was turned into a science by a man named Robert Bakewell during the British Agricultural Revolution of the eighteenth century. Bakewell discovered that he could engineer extra-woolly sheep and beefier cattle by controlling their breeding. As more and more farmers followed Bakewell's lead, farm animals increased in both size and quality. In 1700, the average bull sold for slaughter weighed 370 pounds. By 1786, the average weight had more than doubled to 840 pounds.

As Lohn learned, these same principles are possible with computers using Artificial Intelligence that mimics the idea of natural selection. Just like Bakewell's sheep and cows, an evolutionary algorithm begins with its creator laying out the goals he or she is trying to achieve. "With an antenna, for instance, you might tell the algorithm that you want a solution that will fit in a 4-inch x 4-inch box, be capable of radiating a spherical or hemispherical pattern, and be able to operate at a certain Wi-Fi band," he says. "You provide all the constraints and, based on them, the algorithm then optimizes a solution."

Lohn estimates that it took between thirty and forty cross-country trips from his home in California to NASA's headquarters in Washington, DC, to convince the higher-ups that evolutionary algorithms were the way of the future. Eventually they agreed to give it a go. Lohn got hold of the specs for the then-upcoming Space Technology 5 mission, punched in the antenna's basic requirements and let his software do its thing.

Several hundred generations later, what the algorithm came up with looked like a mistake. Lohn describes the design as resembling a "bent paperclip." He felt deflated, like the person who vouches for his friend to be given a job, only to see them get drunk on the first day and fall asleep at their desk. Nevertheless, Lohn dutifully made a physical prototype and put it in a test chamber. It worked better than any other solution he had ever seen. The same proved true for later designs, although Lohn was at a loss to explain why they worked as well as they did due to the number of superfluous elements they seemed to include.

"Often we will see one or two aspects of the computer's design that we understand as engineers, but the rest of it we don't," he says. "If I optimize an antenna using evolutionary algorithms, there's only a 50 percent chance I'll be able to explain exactly why it has made the choices it has. The rest of the time the design is simply not intelligible to us. It works—and as engineers what we ultimately care about is getting things to work."

Not every solution a genetic algorithm comes up with is so seemingly counterintuitive. Because the fitness function is inputted by the programmer, it is within the algorithm's ability to create variations only on the standard solution the operator is expecting to see. Like Google's Deep Dream project, though, it's equally possible to let the computer go nuts and imagine solutions unencumbered by what has come before.

In Lohn's case, the results spoke for themselves. If his NASA bosses had been skeptical before, they certainly weren't anymore. On September 6, 2013, NASA launched its Lunar Atmosphere and Dust Environment Explorer (LADEE) mission to study the moon's lunar dust environment. On board were three antennae, all of them designed by Lohn's AI. "They were the only antennae on the mission," he says. "If they had failed, there were no other antennae to save the day." The mission was a resounding success. Lohn describes it as a career highlight.

The Lovelace Test

In chapter four, I discussed the Turing Test, which to date remains the most famous test for measuring machine intelligence. However, as the question of creativity has become more prominent within AI circles, some people have suggested that a new test needs to be created.

The first person involved with modern computing to discuss the subject of machine creativity was one of the world's first computer programmers, Ada Lovelace. The daughter of no less a creative force than the Romantic poet Lord Byron, Ada Lovelace worked alongside Charles Babbage on his Analytical Engine in the 1800s. This was intended to be history's first mechanical general-purpose computer, although due to a lack of funding it was never completed. Lovelace was impressed by the idea of building the Analytical Engine, but she argued that it would never be considered capable of true thinking, since it was only able to carry out pre-programmed instructions. "The Analytical Engine has no pretensions whatever to originate anything," she famously wrote. "It can do [only] whatever we know how to order it to perform."

The appropriately named Lovelace Test is a way of testing Ada's theory by asking a computer to come up with a spontaneous creative idea. The Lovelace Test imagines an experiment in which the artificial agent is a, the human creator is h and the original concept is o. The test can only be passed if a is able to generate o without h being able to explain how this has been achieved. To avoid o simply being a random occurrence such as a fluke error, a must also be able to replicate o at the request of whoever is judging the contest.

Researchers hold mixed opinions regarding how close AI is getting to passing the Lovelace Test. For my money, we're getting very close—and we may even be there already. An evolutionary algorithm like the one Jason Lohn created certainly fulfills a number of the criteria. It is an artificial agent able to generate an original design that cannot be explained by Lohn himself, although it can be proven to work effectively.

Lohn's evolutionary algorithm is far from an outlier. In addition to engineering, evolutionary algorithms are also being used in a whole range of areas requiring creativity. For instance, some architects use evolutionary algorithms to come up with thousands of variations on specific styles of building, resulting in designs that look like the Google Deep Dream version of an office block or a Baroque cathedral. Another evolutionary system called "Eurequa," developed by the director of Cornell University's Creative Machines Lab and one of his PhD students, is doing experiments aimed at describing natural laws. Eurequa, creator Hod Lipson says, "is not a passive algorithm that sits back, watching. It *asks questions. That's curiosity.*"

Evolutionary algorithms are good for coming up with creative strategies, too. In 2010, a *StarCraft II* gamer with a background in Artificial Intelligence used an evolutionary algorithm to create a piece of software capable of generating the perfect tactics for the game. Called "Evolution Chamber," the algorithm allowed players to set goals within the game and would then advise on the fastest way to achieve this, by detailing the exact order in which they should build units and structures in the game. It is not difficult to imagine that such creativity might have a place in business, law or medicine. A CEO, lawyer or medical researcher isn't just there to crunch data and tell you to repeat what worked in the past. The most successful examples of all three of these will be able to apply creativity to what they are doing—coming up with a creative new way to solve supply problems, formulate original legal arguments or develop new drugs.

In the summer of 2015, I paid a visit to the University of Manchester's Institute of Biotechnology. Inside an air-conditioned room bearing the somewhat uninviting warning, "Danger of Death: You Must Leave Immediately," I met EVE, a so-called "robot scientist" designed to automate drug discovery. Because developing a new drug can cost upward of $480 million, handing this kind of work over to Artificial Intelligence means more drugs are discovered. It also means that more drugs are discovered for some of the world's most vulnerable people, since pharmaceutical companies typically want to make big profits on whatever drugs they develop, and therefore don't dedicate much time or resources to curing tropical diseases that overwhelmingly affect poor people. EVE (the successor to an earlier robot scientist called, yes, ADAM) not only carries out the testing of new drugs, but also comes up with hypotheses about what to test. As such, it formulates theories to explain what it sees, devises experiments to test these theories, physically carries out experiments and then interprets the results.

"If it were a human being doing this work, it would certainly be considered creative because it's based around formulating and testing hypotheses," says Ross King, EVE's creator and Manchester's Professor of Machine Intelligence. King tells me he is working toward what he calls the automation of science. "In my view, it's only a matter of time before creative machines like EVE become the norm," he opines.

So if drug-discovery robots can be considered creative within the field of medicine, how do they stand as creators in the eyes of the law? According to John Koza, one of the fathers of evolutionary algorithms, this threshold may have already been passed. Like a lot of scientists, John Koza likes his quantifiable proofs. Inventiveness is a good quality to possess, but an inventive idea only drives us forward if it is one that hasn't been seen before. One way of measuring the originality of an idea is whether or not we can patent it. Several years ago, Koza began using evolutionary algorithms as "automated invention machines." So far, his algorithms have created around seventy-six designs able to compete with humans in a range of areas, such as electrical circuit design, optics, software repair, civil engineering and mechanical systems. In most cases, the designs had already been patented by humans. The computer had gotten there too late—although the fact that it came to the same creative solution with no prior knowledge of the work in question shows its value in the design process. In the other instances, Koza's automated invention machines were capable of creating new, patentable concepts.

To Koza, what makes the awarding of a patent so important to the question of machine creativity is a tiny two-word phrase hidden in the US Patents and Trademark Office's legal small print. According to USPTO, a patent can only be awarded to an idea that is deemed to be an "illogical step." "What the patent office means by an 'illogical step' is that it is more than something you would inevitably think of were you to plod along working on something based entirely on the past," Koza explains. Why does that matter? Because, Koza says, if computers are only good at performing logical operations, how can they take an illogical step?

It would certainly be enough to have Ada Lovelace scratching her head.

Congratulations to the Chef

In September 2015, I ate my first meal from a recipe prepared by an AI. It was an Indian turmeric paella, a combination of cuisines that was both unusual and, as it turned out, delicious. I was following a recipe created by IBM's Watson, the *Jeopardy!*-winning AI described in the previous chapter. "The ideas for the recipes in this book weren't generated by your average chef," reads the introduction in a cookbook entitled *Cognitive Cooking with Chef Watson*. "What kind of eccentric would ever dream up a Turkish-Korean Caesar salad or a Cuban lobster bouillabaisse? In this case, it's one that has never tasted a single morsel of food. This culinary prodigy, in fact, has no taste buds, no nose, nor any sensual experience of food or drink."

Perhaps surprisingly, Chef Watson is the project IBM settled on after Watson beat Ken Jennings at *Jeopardy!* As it became evident that Watson was capable of parsing the complex question-as-answer conundrums of *Jeopardy!* in a way that could defeat even the most skilled of humans, employees at IBM Research decided the next logical challenge was to go beyond answers and develop a system capable of being creative. The suggestions rolled in— covering bases like art, literature and music. In the end, Lav Varshney, a scientist for IBM's Smarter Cities initiative, came up with the idea of building an AI that would innovate in an area of almost universal human appreciation: food.

"We started by getting Watson to analyze around 9,000 existing recipes," says Rob High, chief technology officer at Watson Solutions. "From that, the system was able to learn the different types and styles of recipe that exist. It learned the difference between a salad and a sandwich, or a quiche and a pasta dish. It also learned the difference between Vietnamese cooking and Southwestern styles, or French and Chinese cooking. It figured out which flavors come out most prominently within all those types of dishes."

This probably would have been enough to start finding the taste connections that could help it generate Austrian grilled asparagus or Scandinavian salmon quiche (two more of the faintly alliterative recipes on offer), but IBM wanted to drill down further. Watson's researchers decided to fuse the 9,000 recipe data set with another one, describing what Rob High calls "the knowledge of taste chemistry." As High explains it, a human chef may have an encyclopedic mind when it comes to selecting combinations of ingredients, but Watson is able to analyze the chemical compounds that control taste and use these to generate entirely novel pairings. "It goes through something like 6 quintillion permutations* to find the chemical compounds, and the ingredients which contain those compounds, to make you the perfect meal," he says. "Quite often, they're ones you would never expect—although they turn out to be delightful." (One example of a unique food pairing that works shockingly well is cherries and mushrooms. Seriously.)

At present, there are several versions of Chef Watson. As with any celebrity chef, you can buy the official recipe book. Or you can download the smartphone app, which gives users a greater level of control over what the culinary AI chooses. Users start by inputting up to four ingredients they want in the dish, perhaps choosing whatever is hanging around in the fridge. They then choose the style of cuisine they're after, and whether or not there are ingredi-

---

*That's 6,000,000,000,000,000,000.

ents to avoid. Watson immediately processes this information and outputs a recipe, right down to the step-by-step cooking instructions.

It's not hard to imagine it going further than it currently has. Early versions of Chef Watson generated lists of compatible ingredients only, which were then turned over to human chefs to interpret into recipes. Newer versions now create the recipes themselves, right down to advising on quantities and cooking times. Within the next decade it's not inconceivable that we could have a kitchen robot, not unlike the Momentum Machines burger-making robot described in the last chapter, capable of carrying out the whole process end-to-end. Find yourself having a sudden craving for a Korean-inspired chicken quiche, regardless of whether or not it exists? No problem. You could just set the parameters from your iPhone 12 as you leave the office, and have the food ready for eating as you arrive home.

But Rob High and the other IBM employees I spoke to were keen to stress that Chef Watson isn't the end goal for IBM's Artificial Intelligence. Like getting one of the world's most powerful AIs to compete on a game show, at its root, Chef Watson is a metaphor—a proof of concept to show off the way Watson can use its enormous database of natural-language knowledge to work in a variety of areas. Creativity isn't just about preparing new dishes, but coming up with ways to aid humans across industries. For instance, one area IBM is working in is medicine, with Watson helping oncologists choose more personalized, creative paths for curing cancer. Another area is the legal profession, with Watson helping to find case law, which may help shape more creative arguments on the part of lawyers. Another could be business analytics.

"However smart these systems get, I think it's important to see these developments as tools," says Rob High. "Something like Chef Watson isn't about supplanting human creativity; it's about amplifying human creativity. It's going to allow us to be more creative than we could ever be without it. That goes way beyond cooking. It's a twenty-first-century paintbrush."

A paintbrush that will increasingly paint its own pictures, that is.

7

In the Future There Will Be Mindclones

MARIUS URSACHE WANTS you to live forever.

It's not a shock to find out that, in an industry that skews as young as tech, few people spend much time thinking about death. This is, after all, a walk of life in which twenty-one-year-olds are already onto their second startup, billionaires are minted by twenty-five and even Steve Jobs once fretted about whether people older than thirty were capable of achieving anything of lasting significance. As a result, the idea of growing old and dying is, for most Silicon Valley denizens, the furthest thing from their mind.

As a former medic from Romania, Ursache thinks about death more than most people. He has even turned it into a job. As the creator and founder of a startup called Eterni.me, he spends his days working toward the dream of building Artificially Intelligent 3-D avatars: digital beings that will look, sound and, most important of all, act like individuals who are no longer with us.

Ursache's journey began several years ago when he became fas-

cinated by the game *Second Life*, a vast online virtual world created by the San Francisco-based developers Linden Labs. Although
Second Life resembles a computer game, it differs in one crucial
sense. Rather than featuring set objectives and manufactured sto-
rylines, players in *Second Life* refer to themselves as "residents" of
the game, and participate in any way that they wish, whether that
means running a shop, or simply hanging out with friends.

"One day, I started wondering what happens to a person's avatar in the game after they die," Ursache says. Was there, he
pondered, a kind of *Second Life* purgatory where abandoned avatars lived on in a zombie-like state, long after their human operators had passed away? What would happen if one tried to interact
with these avatars?

He began to consider the idea more and more. He attempted to
work out the logistics of programming an artificial agent that
could convincingly mimic the behavior of its human counterpart.
He thought about the kind of code one would need to write for an
avatar so that it could learn to move the way its human player once
moved, to talk the way that they once talked, and to form and pursue the kind of goals they might have created and pursued. And as
with any enterprising entrepreneur, he tried to think of a way to
turn it into an actual product.

In February 2014, Ursache was invited to attend a program for
entrepreneurs at MIT, by a mentor he had met in Bucharest. As
part of the program, he was asked to come up with an idea for a
project to work on. By this time, the idea had expanded in his
mind. "What I was thinking was that this could be a great way of
letting you collect and curate your digital footprint throughout
your life," he says. "The avatar would be an interface for accessing
that information."

He pitched the idea to the group as "Skyping with dead people," and hurried to note that a lot of the AI technology needed to bring such a project to life already existed in various labs around the world. Despite the group receiving a total of 130 ideas—of which Ursache acknowledges his was the oddest—"Skyping with dead people" was accepted as a project worth pursuing.

Ursache had his reservations, however. "I knew that it would have to do something more than just simulating a conversation with a dead person," he says. "That would be too fucked up. It would mess with the grieving process and, frankly, would just be weird."

He decided to put up a webpage to gauge the reaction of the general public. If they responded positively, he would keep working on it. If the idea met a wall of indifference, or even anger, he would drop it.

Within the first four days, the page had 3,000 people sign up to register their interest. That number quickly rose to 22,000, and then kept right on climbing. There were plenty of messages, too, which Ursache dutifully read as a form of market research. Most of them were full of praise for the project, although a certain percentage (he estimates around one-fifth) talked about how creepy it all sounded. Who wanted a version of Siri that acted and spoke like their dead grandparent?

Then Ursache received the e-mail that changed his life. It was from a person dying of terminal cancer. In their e-mail they explained that they had six months left to live. A project like Eterni .me, they wrote, was their chance to leave something behind for friends and family.

"It was easy to reply to the messages from people who were congratulating or criticizing us," Ursache says. "But what could I say to someone who was dying? That was the moment I decided that this was something worth dedicating my life to." Almost overnight, Ursache made the decision to pack in his previous job and focus on Eterni.me full-time.

Today, Eterni.me has 30,269 eager subscribers, all waiting on their ticket to digital immortality. The company's website shows video clips depicting a range of memories from the average lifetime. A bride and groom kiss on their wedding day. A mother hugs her baby. A child plays at being a superhero in the garden. Graduating students throw their caps in the air. Retirees laugh together. "What if . . . you could preserve your parents' memories forever?" Ursache's marketing blurb reads. "And you could keep their stories alive, for your children, grandchildren and for many generations to come? What if . . . you could preserve your legacy for the future? And in this way your children, friends, or even total strangers from a distant future will remember you in a hundred years? What if . . . you could live on forever as a digital avatar? And people in the future could actually interact with your memories, stories and ideas, almost as if they were talking to you? Eterni.me collects your thoughts, stories and memories, curates them and creates an intelligent avatar that looks like you. This avatar will live forever and allow other people in the future to access your memories."

Currently the technology doesn't exist to allow us to "Skype with dead people" as Ursache would eventually like. While his team work on the machine-learning tools that will make the technology a reality, Eterni.me instead focuses on collecting the users' data that will one day give its avatars their digital lifeblood. He doesn't think Eterni.me's 30,269 early adopters are going to be waiting forever, though.

"This isn't technology that is decades away," he says. "Building lifelike avatars is an iterative process. Think of it like search results; they'll just get better and better, more and more accurate as time goes on."

Yourself in Machine Form

As it happens, Marius Ursache is far from the first person to consider how machines might allow humans to live on after their death. Despite only dating back a few decades, multiplayer online games have already had to grapple with what happens when a popular player or creator dies. In *Dungeons and Dragons Online*, for instance, the death of original tabletop game cocreator Gary Gygax resulted in the creation of an in-game mission narrated by his recorded voice. It is a disembodied voice from beyond the grave—a reminder of which is the fact that Gygax's virtual tomb can be discovered in the game close to where the mission takes place.

There are other, more common examples of how humans and their decision-making process can become embedded in machine code, though. The "expert systems" described in chapter one were, in essence, attempts to create clones of flesh-and-blood human experts. This was done by extracting their specialized knowledge and turning it into a set of probabilistic rules, capable of processing by a computer. Had expert systems worked out as planned, a particularly brilliant cardiologist, attorney or CEO could have their expertise reproduced and distributed throughout the world. Much like laws continue to be followed after lawmakers have passed away, the idea of an expert system is that we ought to be able to continue drawing on an expert's knowledge about a specialized subject after the person is no longer available to us. The concept failed, but the intention (and, for a while, the funding) was absolutely there.

In some senses, the modern parallel of the expert system is the so-called "recommender system." This subclass of information filtering system sets out to anticipate and predict what rating or selection a user is likely to give an item in a specific narrow domain. Everyone reading this will likely have come across the feature on Amazon or Netflix that suggests that, "You liked X, so you may also enjoy Y." Sometimes these predictions are less than stellar, but as much as we like to think of ourselves as fundamentally unpredictable beings, it's often surprising just how accurate they are.

Companies certainly think so, at least. In January 2014, Amazon was granted a patent for a shipping system designed to slash delivery times by predicting what people are going to want to buy before they buy it and sending the item out early. In some cases, Amazon explained that it might send out products to a local shipping hub until the order rolls in as expected. In other situations, it noted that it could send out targeted promotional gifts in order to establish customer goodwill.

As with so many of the other technologies described in this book, the key to this kind of prediction is the ability to analyze enormous amounts of user data. Every year we save more bytes of data about ourselves than we have base pairs of DNA. Each time we enter cyberspace we leave traces of ourselves behind. Whenever we blog on WordPress or LiveJournal, post a new status update on Twitter or Facebook, comment on the news using Contextly, choose a movie or TV show to watch on Netflix, send IMs or simply make searches with Google, our digital identity is updated and curated. The result is an increasingly accurate picture of who we are, represented in ones and zeroes. Over the coming years, this once-fuzzy outline will become more akin to a detailed line drawing or even a photo-realistic painting as more and more of who we are as people becomes discoverable to AI systems. As we saw in chapter three, this knowledge will allow the world to re-configure and optimize around us. The door to your home or your hotel room will open for you and only you when you approach it. Rental cars you've never driven before will automatically adjust to meet your preferred settings. Thermostats will know how hot or cool you like the temperature and adjust themselves to enhance your productivity at different times of day.

This isn't just a case of more data being gathered, but new and different types of fine-grain data that are routinely recorded by hundreds of millions of people. For instance, the Apple Watch collects your heart rate over the course of your life, while various Apple Watch apps can cross-reference this information with a variety of other data points. At the 2014 Consumer Electronics Show in Las Vegas, Sony meanwhile introduced its "life logging" software, designed to track people's activities—from the phone calls they make to friends to the movies they watch—on an interactive timeline. Both of these data sources were unimaginable in the 1990s as anything other than one-off experiments in research labs.

Personality Capture

For most of William Sims Bainbridge's life, the seventy-six-year-old codirector of Cyber-Human Systems at the National Science Foundation has been obsessed with the subject of life after death.

In his office, he keeps a photograph of his great-aunt Cleora, which he says he does to remind him of his own mortality. The photograph is old, taken in 1870, and the emulsion is peeling off its ancient metal backing. It shows Cleora Emily Bainbridge at the age of one, shortly before her untimely death. While many of us have ancestors who died at a young age, there is something very unusual about what happened to Cleora after her passing. When she died, her clergyman father decided to memorialize her by imagining the life that she never got to live. He did this in the form of a novel, written in 1883, which he titled *Self-Giving*. In the story, Cleora grows up to become a Christian missionary, before dying a martyr.

Whatever led to Cleora's father taking this unusual step concerning posthumous life preservation, it seems to have been hereditary. Almost a century later, in 1965, Bainbridge's parents and sister were tragically killed in a house fire. He was in his midtwenties at the time, a sociology student at Harvard. Looking for a way to preserve their memory, the bereaved Bainbridge spent months sifting through the ashes to salvage whatever memorabilia he could find.

This interest in preserving memories eventually led him to neural networks, which in the early 1980s had just proven capable of storing associative memories thanks to the work of John Hopfield and others. Inspired by what he saw, Bainbridge imagined that large enough neural networks may one day be able to store the sum total of a person's memories. He later called this idea "personality capture" after the technique known as "motion capture," in which a person's movements are scanned into a computer for use in video games or movies. Bainbridge's big idea about personality capture is that it should be possible to accurately model how a person would behave in any given situation by asking people a large number of questions about themselves and then using this information to model something called an AI "mindfile." To create a truly accurate picture of who an individual is would, he suggests, require them to answer approximately 100,000 questions. Each question would then be weighted in two dimensions, with both the relative importance of each personality attribute to the individual in question, and also the relative degree of applicability to that individual. The result would be a software avatar able to respond to real-life situations in the same way as the person that it is based on.

Although this has not been done yet, there is no obvious reason why it should not be possible. In the 1980s, a researcher at Bell Communications research called Thomas Landauer carried out a series of experiments to find out how much people remember over the course of a lifetime. These experiments included asking people to look at pictures and hear words, sentences and short passages of music. After delays that ranged from several minutes to several days, Landauer then tested participants on how much they could recall. This was often done using multiple-choice questions. Although his estimates fail to take into account everything, he concluded that the typical person stores roughly two bits of information per second. Over the course of a lifetime this works out at 10^9 bits, or some hundreds of megabytes. Given that the computer that I am writing this book on has three terabytes of storage, according to Landauer's calculations, even a low-end personal computer could store thousands of people's mindfiles.

Much as an artist paints a self-portrait, training a mindfile and subsequent virtual avatar to be exactly like you could be a form of self-expression a person could pursue throughout life. Since most of us are not always conscious of our failings, Bainbridge says that it would also likely be necessary to call in a person's family, friends and associates to provide details about them in areas like trustworthiness, work ethic and the like.

Similar to Marius Ursache, Bainbridge's personality capture would be a way to use AI to model a loved one, or other significant person, so that their memories and opinions can be accessed after their death. The resultant avatars will become interactive history artefacts for future generations, much as we might today regard an old photograph or listen to a recording made of a person while they were alive. You can think of it as a library that stocks people instead of books. A more basic variation on this concept was put into action several years ago by Carnegie Mellon University, for what it called the "Synthetic Interview." The idea of the Synthetic Interview was to bring to life the writings of historical figures like Abraham Lincoln by creating an interactive video experience capable of responding to questions. For the project, actors were dressed up as the historical figures and delivered responses to hundreds of possible questions users might ask. A very basic piece of software then selects the appropriate video in response to each question. "A Synthetic Interview deeply engages people with a person or time period in a way that just isn't possible when passively viewing a film," says Don Marinelli, executive producer at Carnegie Mellon's Entertainment Technology Center. "And when we pair this technology with a figure as revered as Abraham Lincoln, the effect is powerful."

Ultimately, personality capture has the potential to become the next step in a sequence which began when prehistoric societies first painted images on the walls of caves as a symbolic means of overcoming death. When we look at the *Mona Lisa* hanging in the

Louvre, or we read *Macbeth*, we get a glimpse of the mind of the person who created them, despite the fact that Leonardo da Vinci and William Shakespeare have been dead for hundreds of years. A so-called "mindclone" might one day do the same. In addition to recorded memories, AI behavioral modeling could even be used to predict how a person might have responded to events taking place after their own life—like how it is possible today to access the Amazon account of a deceased relative, provided that they used the services, and be recommended books, films and music albums to their taste, some of which may have been created after their death.

Of course, the kind of "virtual immortality" that Marius Ursache talks about is a very different kind of immortality to the one that means living on as a conscious being. As Woody Allen once quipped: "I don't want to achieve immortality through my work; I want to achieve immortality through not dying."

Could Artificial Intelligence help with this, too?

Robots Will Inherit the Earth

The possibility of one day using science to preserve the mind by transferring it beyond the limits of the human body is one that has been speculated upon for many years. It can be traced back to an influential Russian philosopher named Nikolai Fedorovich Fedorov. Born in 1829, Fedorov was a member of the Russian Cosmists: a movement that fused together a smörgåsbord of Eastern and Western philosophies, with a religiosity straight from the Russian Orthodox Church. The quasi-religious zeal of some of today's futurists—usually sat side by side with the latest cutting-edge science—owes a lot to the legacy of Russian Cosmism.

Fedorov imagined evolution as a process centered around intelligence and our quest to achieve it. Man, he argued, was the culmination of natural history, and should use whatever reason and morality it has available to guide the hand of natural selection. This led to an interest in the use of scientific methods to bring about not just radical life extension and physical immortality, but also the idea that we might eventually restore to life everyone who has ever died. Where would all of these newly immortal beings live? In space and under the sea, of course, which Fedorov envisioned being colonized by the human race. Like Google's democratized, data-driven approach to life as we know it, Fedorov stated: "Everyone must be learning and everything be the subject of knowledge and action."

Nikolai Fedorov's ideas were quickly followed by others. The French social scientist Jean Finot wrote 1909's *The Philosophy of Long Life*, in which he advocated the use of science for engineering life and the fabrication of living matter. The British evolutionary biologist J. B. S. Haldane's 1924 *Daedalus; or, Science and the Future* later argued that evolution could be carried out in the womb by way of directed mutation. Haldane's ideas were a major influence on the author Aldous Huxley when he wrote his dystopian novel, *Brave New World*. The 1927 book *The Struggle for Viability* next saw Bolshevik writer Alexander Bogdanov claim that natural lifespans were being artificially shortened by social imperfections. Should these be improved, Bogdanov wrote that he perceived no reason why our lives should not "last 120 to 140 years" at the very least. X-ray pioneer J. D. Bernal meanwhile penned *The World, The Flesh and the Devil* in 1929, elaborating on Fedorov's concepts of space colonization and describing his theories on the enhancement of both human lifespan and intelligence.

These screeds make fascinating reading today, but it was not really until computers, robotics and Artificial Intelligence became a part of the conversation in the late twentieth century that the dreams of "transhumanists" really took hold. Particularly as they grew older, many of the old guard of Artificial Intelligence turned their attention to technology that would help them live long past their natural lifespans.

One such person was Marvin Minsky. In 1994, at the age of sixty-seven, Minsky wrote an article for *Scientific American*, entitled "Will Robots Inherit the Earth?" In his 4,500-word essay, Minsky made the case that we are well on our way to solving the problem of death as we know it. He divided death into two separate fields: the breakdown of the body and the breakdown of the mind. Bodies, he suggested, were the (relatively) easy part. Already, he noted, average human lifespans are up from the twenty-two years they were in ancient Rome, and the half century they were in developed countries in the year 1900. As science continues to learn more about genetics, Minsky argued that we will be able to postpone many of the conditions that currently affect people in their latter years. Drawing a parallel with the animal kingdom, he pointed to an experiment involving a certain type of Mediterranean octopus. After spawning, this species of octopus stops eating and soon starves to death. However, if scientists remove the optic glands of females brooding their eggs, they discovered that the octopus abandons its eggs, resumes feeding and growing and survives for roughly twice the length of an unmodified octopus. Already we carry out not dissimilar operations on humans for the purposes of life extension, whether it be removing dangerous tumors or giving people potentially lifesaving transplants of vital organs.

As one might expect from a person who had spent their career investigating intelligence, Minsky was of the opinion that minds were more complicated. He wrote:

> As a species we seem to have reached a plateau in our intellectual development. There's no sign that we're getting smarter. Was Albert Einstein a better scientist than Newton or Archimedes? Has any playwright in recent years topped Shakespeare or Euripides? We have learned a lot in 2,000 years, yet much ancient wisdom still seems sound—which makes me suspect that we haven't been making much progress. We still don't know how to deal with conflicts between individual goals and global interests. We are so bad at making important decisions that, whenever we can, we leave to chance what we are unsure about.

The idea of creating a "hive mind" of sorts captured the zeitgeist at a time when the Internet was just taking off. But Minsky didn't simply want to save information and make it available for future generations. He wanted to be around to see it. "Eventually we will entirely replace our brains using nanotechnology," he wrote. "Once delivered from the limitations of biology, we will be able to decide the length of our lives—with the option of immortality—and choose among other, unimagined capabilities as well."

The Connectome

A complex recommender system "mindfile" of the sort described by Marius Ursache and William Sims Bainbridge may go some way toward replicating us in software form. However, the only truly faithful means of making sure that a person is reconstructed in a form other than their original one would be to duplicate all of the cellular pathways in the brain—neuron by painstaking neuron.

For this to be possible, we must first accept the central tenet of Artificial Intelligence: that the main task that the brain carries out can be viewed as information processing not dissimilar to that which is carried out by a computer. In other words, that there's no substantial difference between the software used in a computer system and what is sometimes referred to as the "wetware" of the human brain. This model of intelligence asks us to go along with the principle of "substrate independence," meaning that the brain as a dynamic process is not specifically tied to a set of atoms. If the brain's information processing really is substrate independent, then this means that one day it will be possible to transfer intelligence from a protein-based brain to another, more durable medium such as a computer network.

The question therefore becomes how we construct such a brain. The simplest answer, as anyone who has ever taken apart an alarm clock to see how it works will know, is to "reverse-engineer" it. This is the act of taking apart a piece of existing software or hardware in order to understand how it is made. Once we know how it is made, we can then build an identical model in the same way that we could train a neural network to behave identically to any described in this book if we happen to know how it has been constructed and have access to the same inputs and outputs.

Today's most successful attempts to model the brain as software are deep learning neural networks. From simple metaphors for the biological brain, these networks have grown increasingly complex and are getting better all the time. When Marvin Minsky wrote his 1994 essay, "Will Robots Inherit the Earth?," a large neural network consisted of around 440 connections. As I write this chapter, the world's largest deep learning network belongs to a US cognitive computing company called Digital Reasoning, and features around 160 billion neural connections. This is a significant leap in just a couple of decades, although it is still far away from the actual complexity of the human brain, which is home to approximately 86 *trillion* synaptic connections. Per cubic millimeter of human brain tissue, there are an astonishing 100,000 neurons and roughly 900,000,000 synaptic connections.

Provided that Moore's Law holds out,* building a neural network this size is not out of the question within the coming decades. Unfortunately, this alone will not be enough to generate brain-like intelligence. We know this because computer scientists have already built neural networks with well over a million neurons and yet still don't have general purpose AI on the level of a comparable animal. In the animal kingdom, 1 million neurons should give us a brain with the intelligence level of a honey bee (960,000 neurons) or a cockroach (1 million neurons). We have not yet done this. In fact, the closest we have got to re-creating the "connectome," or wiring diagram of the central nervous system, for a real animal is the work done analyzing the tiny hermaphroditic roundworm called *Caenorhabditis elegans*. The Nobel Prize–winning biologist Sydney Brenner and his colleagues began slicing up *C. elegans* back in the 1970s in order that his team could photograph

---

*Moore's Law, named after the cofounder of Intel, Gordon E. Moore, holds that the overall processing power for computers will double approximately every year (later revised to every two years). Moore first made this observation in 1965 and so far it has held up remarkably accurately.

the cells using a powerful electron microscope. By 1986, Brenner had gathered enough information to publish a connectome of the creature's complete nervous system. It remains the only full connectome of any living creature that we have been able to decipher.

As systems go, *C. elegans* is fairly basic—with just 302 neurons connected together by 7,000 synapses. But despite this comparative simplicity, we still only have the slightest understanding of how its nervous system actually works. Since 2011, the task of modeling *C. elegans* inside a computer has been worked on by an international collaboration of hundreds of scientists and programmers in the United States, Europe and Russia. Called OpenWorm, the project has so far managed to construct both a simulated physical body model of the worm and a detailed simulatable model of the creature's nervous system. However, even with hundreds of thousands of man-hours, we still don't yet have enough knowledge about how *C. elegans*' neurons process information that we can reproduce even the most basic of crawling behaviors.

The reality is that it is no longer a big deal to be able to simulate a billion neurons interacting with one another, provided that we have access to a powerful enough computer. From the work that has been carried out, we know that assembling billions of neurons in a network won't result in an intelligent, human-level brain any more than putting a billion transistors together will result in a functioning central processing unit (CPU). A connectome like *C. elegans* is a circuit at rest, lacking virtually all of the information about the way in which the circuit operates. This is because there are parameters hidden inside the neurons we simply don't have access to when we look at the network. To put it another way, you might be able to build a computer by looking at the blueprints of an existing one, but you would remain nonplussed about how to program Microsoft Word.

If the mind is the software of the brain, why would we expect it to be any different?

Mapping the Mind

Up until now, neuroscience has mainly taken place at two ends of the spectrum of scale. At one end are researchers who are focused on the micro study of individual neurons. This has led to some advances, but it provides limited knowledge about the way the brain functions because it ignores the network activity of the brain that is going on around the neurons. At the other end are those who are interested in the macro-cortical architecture of the different parts of the brain, at which the smallest resolvable unit may be hundreds of thousands of neurons. This type of study was traditionally carried out by physically removing parts of a person's brain and analyzing it under a microscope. Today, it is possible to do this in less invasive ways. In 1990, the Japanese physicist Seiji Ogawa and his colleagues created a brain imaging technology called functional magnetic resonance imaging, or fMRI. By working out which parts of the brain are responsible for certain types of behavior, remarkable things are starting to be achieved.

In 2015, for instance, doctors in California implanted twin electrodes in the brain of a 34-year-old quadriplegic named Erik Sorto, allowing him to control a robotic arm using only his thoughts. By recording signals in Sorto's posterior parietal cortex (the part of the brain that deals with movement planning) and feeding these into a neural network designed to analyze the signal,

Sorto's intentions were able to be decoded and then translated into movement commands for the freestanding robotic arm. Sorto began with simple tasks like shaking hands, but was soon able to graduate to play rock, paper, scissors, and even pick up and drink a beer by himself for the first time in over a decade.

In another similar experiment, University of Houston researchers developed a brain-machine interface that required no brain implant and only an electroencephalogram (EEG) brain cap, through which it could detect the brain's electrical activity through the scalp. Despite the resulting signal being "noisier" than the one you would get from physically placing a nanoelectrode inside the brain, researchers were able to narrow down and amplify the frequencies that useful brain signals operate on. As with Erik Sorto, the result was that the fifty-six-year-old amputee used in the test was able to pick up various objects, including a water bottle and a credit card, using a robotic hand.

As with the detail of the "mindfile" described at the start of this chapter, knowledge extracted from projects like these are enabling researchers to build up ever more detailed pictures of the brain. In 2013, a team of researchers produced a 3-D human brain scan that takes up one terabyte of space (1,000 gigabytes). Scans like these tell us more than ever about details such as the brain's microanatomy, although they're still not sufficient to answer questions about its microstructure.

The next step will therefore involve drilling down in even more granular detail to work out how things function on a neuron-by-neuron scale. Right now, neuroscientists know roughly what it is that neurons do in the brain, and how it is that they communicate with other neurons, but they are unable to decisively say exactly what it is that is being communicated. Neurons come in hundreds, possibly even thousands, of variations; each with their own cell types and unique molecular identities. At present, we still do not know how many different classes of neuron there are, or what the electrical or structural properties are of each type. Nor do we know what format the brain uses to encode, in the way that we understand that computers use file formats such as JPEG and GIF to encode images, or DOC and TXT for text documents. Understanding the brain means not simply understanding how individual neurons work, but also how they interact with other neurons in parallel as part of a network.

There are different ideas about how this is best achieved. Futurist Ray Kurzweil, currently employed at Google as one of its directors of engineering, has suggested using tiny microscopic nanobots to scan the brain. A bit like the injectable smart devices described in chapter three, Kurzweil's vision calls for billions of these scanner nanobots, the size of human blood cells or even tinier, to enter the brain and capture "every salient neural detail" by scanning from inside.

It's a good idea in theory, but Kurzweil's optimistic proposal has been criticized by some neuroscientists for being the brain science version of suggesting that we raise awareness of endangered species by building a massive highway through the middle of the rainforest so that people can get a closer look at the animals. For instance, David J. Linden, chief editor of the *Journal of Neurophysiology*, points out that the brain is not composed simply of neurons, but also of what are called "glial cells," which outnumber neurons ten-to-one and are packed together far too tightly to allow a nanobot through. Making things worse is the fact that even the minuscule space between brain cells is filled with support structures used for ferrying signals back and forth to neighboring cells.

"You can imagine Kurzweil's brain nanobot . . . crashing through this delicate web of living, electrically active connections," says Linden. "Even if our intrepid nanobot were jet-powered and equipped with a powerful cutting laser, how would it move through the brain and not leave a trail of destruction in its wake?"

But if Kurzweil is wrong about some parts of the story, he's not wrong about the bigger picture. Along with advances in Artificial Intelligence, parallel developments in nanotechnology, robotics and neuroscience is the reason why billions of dollars are currently flowing to support reverse-engineering the human brain. Like the 1956 Dartmouth conference that kick-started AI, this is resulting in some fascinating collaborations between disciplines.

The Next Big Thing

One such project is the Machine Intelligence from Cortical Networks project, also known as MICrONS. Funded by IARPA, the United States' Intelligence Advanced Research Projects Activity department, the goal of MICrONS is to increase machine intelligence by building algorithms that function more like the human brain. The advantage of this is that, while computers are much better than humans in particular contexts, the human brain can still perform other tasks much more effectively than machines. For example, we are far superior than even the most advanced neural network at generalizing based on a small data set. It may be possible for a neural network to beat the human brain at certain visual recognition tasks, but in order for it to do so a computer needs to see thousands—or even millions—of training examples. A human, on the other hand, may happen to see thousands or millions of one particular object over the course of a lifetime, but they do not need to see all of these in order to recognize that object. If you are shown a device or an animal you have never seen before, you will likely need to see it only a handful of times in order to be able to pick it out of an assortment of other objects, regardless of the angle or lighting conditions it is shown to you under. This is because neural networks remain brain-inspired technologies, not an actual re-creation of the brain.

"Most of today's state-of-the-art algorithms are derived, at least at a high level, from neuroscience principles," says MICrONS' project lead, R. Jacob Vogelstein, an expert in biomedical engineering. "But those neuroscience principles are now twenty, thirty, in some cases fifty years old. There really hasn't been a lot of technology transfer between the neuroscience and machine learning communities in many decades." What Vogelstein says he wants to do is "close the gap" between current AI algorithms and the ones actually found in the brain.

The MICrONS project calls for experimental neuroscientists to do high-resolution structural and functional imaging of the brain, applied mathematicians to analyze brain "graphs," computational neuroscientists to model neurons and neural circuits, and—finally—machine learners to use this data to build algorithms that exhibit more humanlike characteristics.

Two similarly large-scale research projects currently under way are the Brain Research Through Advancing Innovative Neurotechnologies project and the European Commission's Human Brain Project. Carrying the somewhat-redundant backronym BRAIN, the first of these initiatives was announced by President Barack Obama at his State of the Union address in early 2013. Its goal is to map the brain at the level of its electrical pathways and, in doing so, to shed light on various neurological disorders, such as Alzheimer's, Parkinson's disease, schizophrenia, depression and traumatic brain injuries.

The $1 billion Human Brain Project, meanwhile, has the stated aim of building a complete computer simulation of the human brain over a ten-year period in Geneva, Switzerland. (Perhaps it will be the brain of a banker?) To achieve this, its directors plan to reverse-engineer the brains of various animals in order of complexity, starting with a mouse and working their way up. There are multiple prospective advantages to a project such as this. A complete simulation of the human brain would enable the development of superior diagnosis and medical treatment tools, since their impact could be rapidly and easily tested on the artificial brain model. A better understanding of how the brain works could also lead to a revolution in fast, energy-efficient computing: thereby taking technologies such as data mining and telecommunications to the next level.

Because of the collaborative, government-led nature of all these projects, comparisons have been drawn with other significant research projects from the past century, including the Human Genome Project, man's voyage to the moon, and the development of the atomic bomb. While all of them have different approaches and goals, it is hoped that they will all increase our understanding of how neurons are wired and, more important, how they function together dynamically.

Not all such projects are taking place in the public sector, however. One of the more unusual and ambitious brain-science projects of the past few decades was announced by Russian billionaire Dmitry Itskov in 2011. Given the name the 2045 Initiative, Itskov's project is a nonprofit dedicated to the goal of life extension. In its

own words, the 2045 Initiative aims to "create technologies enabling the transfer of an individual's personality to a more advanced non-biological carrier, and extending life, including to the
point of immortality."

Like the BRAIN initiative, Itskov's project breaks down into
multiple stages. Stage one calls for the building of robots capable
of being controlled by the human mind. Stage two is the development of robots that can host a physical human brain, installed by
way of surgical transplant. A decade after this, Itskov plans to be
able to upload the contents of the human brain into a robot, which
means reverse-engineering the brain. Finally, by the year 2045,
Itskov plans to replace these robots with holograms.

In an open letter to some of the world's richest people, Itskov
pleaded with individuals to back his project—and perhaps even to
consider volunteering as early test subjects. "I urge you to take
note of the vital importance of funding scientific development in
the field of cybernetic immortality and the artificial body," he
wrote. "Such research has the potential to free you, as well as the
majority of all people on our planet, from disease, old age and even
death."

The 2045 Initiative received a great deal of support. But one of
the names on the list of participants intrigued me more than the
others. It was that of a serious, accomplished neuroscientist named
Ken Hayworth.

Is There Life on Mars?

As a young boy, Ken Hayworth wanted to go into space. Like many
kids who grew up in the years following the Apollo missions, he

was fascinated by the idea of space travel and its myriad possibilities for mankind. Even then, Hayworth was ambitious. He didn't just want to visit another planet in our solar system; he wanted to build a new type of rocket ship that could take us to the nearest star.

"I was kind of a nerd about it," he admits. "I delved into the physics of how it might be done. I looked at the different designs people had come up with beforehand, and just read up on it like crazy." He came away from his research discouraged. As far as the young Hayworth could tell, the engineering needed to take humans to the nearest star within even the upper limits of a single lifespan was beyond our wildest technological dreams. Even if he took off tomorrow, the then-teenager would die an old man before getting close to his destination.

"At that point I started thinking about how else this might be done," he says. His reading led him to some books about neural networks. "It dawned upon me that the problem with space flight is really about the problem with human beings. If we could extract a person's mind out of this heavy body that needs life support, radiation shielding and everything else, then that information could be transmitted at the speed of light via radio waves." Almost overnight, Hayworth's focus switched from the hard engineering of building heavy spaceships to the question of how the mind works—and how, if at all, it could be taken apart and pieced back together again.

One computer science undergraduate degree and a neuroscience PhD later and Ken Hayworth is today the president and cofounder of a group called the Brain Preservation Foundation, a faintly Hammer Horror–sounding operation whose creaky name evokes old movies with Peter Cushing and Christopher Lee. (On the subject of names, his cofounder at the Brain Preservation Foundation is a man called John Smart, which is possibly the most perfect name for a person working in his position this side of former Google.org director Larry Brilliant.)

Hayworth speaks about consciousness uploads in no uncertain terms. "I absolutely believe that mind uploading is possible and I think it's something we should actively be working toward," he says. "At the very least, doctors should be giving people the ability to preserve their brains in a high-fidelity manner. That way, when the technology comes along that can scan a whole brain at the synaptic level, it'll be possible to bring someone back with simulated brains in robotic bodies." He pauses before adding, "I'm certainly looking forward to that myself."

He also dismisses the kind of defeatist talk that would have you believe such things are impossible. Would you, he asks in one published paper, be willing to undertake an untested surgical procedure that lowers your body and brain temperature down to 10° Celsius and stops your heart and blood flow for a full hour? During this time, your brain would be non-active since, at 10° Celsius, all communication between neurons is halted—meaning that you fulfill almost all of the legal requirements for death for one full hour before being brought back to life. If you answered the question by saying "no" you would be making a terrible mistake. The "untested" procedure Hayworth describes is actually a real surgical technique called Profound Hypothermia and Circulatory Arrest, used for treating things like difficult-to-reach brain aneurysms. "The only part of this scenario that is unrealistic is the doctor letting his patient commit suicide [by refusing the surgery] over such a flimsy philosophical argument," Hayworth says. "A doctor today would simply point to the hundreds of reports of patients leading high-functioning lives after undergoing the procedure."

Where we are today, compared to where we need to be for consciousness uploads, is the equivalent of where researchers like Frank Rosenblatt were with neural networks in the early 1980s. In this case, the big picture is correct, but we need to keep on plugging. Fresh insights from fields like neuroscience will help, while ever-increasing data sets and Moore's Law will do the rest. In this way, working toward achieving consciousness in a machine is a little like the way Google is perfecting their search engine. Larry Page and Sergey Brin began at Stanford with their PageRank algorithm, which remains the kernel of the Google empire. PageRank ranked pages according to the quality and number of incoming links to each page. But while PageRank remains a crucially important algorithm, Google has since enhanced it with 200 different unique signals, or what it refers to as "clues," which make informed guesses about what it is that users are looking for. As Google engineers explain, "These signals include things like the terms on websites, the freshness of content [and] your region," in addition to PageRank. Human consciousness could well reside in a similar number of clues: a combination of life's training data and millions of years of evolution, which we call instinct.

But we're getting closer. Recent studies have suggested that growing children show signs of the same kind of probabalistic decision-making that drives many of today's AI systems. With billions of dollars in funding, and smart researchers like Ken Hayworth on the job, it is only a matter of time before consciousness uploads are a reality.

Or so he hopes.

"Inventing mind uploading is the equivalent of inventing penicillin," Hayworth tells me, as our conversation comes to an end. "It's something that needs to be done and, once it is, everyone will realize that it was the right thing to do. We won't be able to fathom how we ever lived before it happened."

8

The Future (Risks) of
Thinking Machines

IN NOVEMBER 2014, Elon Musk, the then-forty-three-year-old CEO of Tesla Motors and SpaceX, posted an online comment on the futurology website Edge.org. "The pace of progress in Artificial Intelligence ... is incredibly fast," he wrote. "Unless you have direct exposure to groups like DeepMind, you have no idea how fast—it is growing at a pace close to exponential. The risk of something seriously dangerous happening is in the five-year timeframe. Ten years at most. This is not a case of crying wolf about something I don't understand. I am not alone in thinking we should be worried. The leading AI companies have taken great steps to ensure safety. They recognize the danger, but believe that they can shape and control the digital superintelligences and prevent bad ones from escaping into the Internet. That remains to be seen. . . . "

Just minutes later, he deleted the message. Musk is a lot of things (many of them good), but uninformed he is not. Over the past several years, the electric car entrepreneur has made several investments in machine intelligence companies, including the aforementioned DeepMind—the deep learning company whose work is described at the start of chapter two. With a personal net worth in the region of $11.2 billion, Musk says that he doesn't make his AI investments with an eye on making a return on his investment, so much as he does to stay clued in. "I like to just keep an eye on what's going on with Artificial Intelligence," he told the US news channel *CNBC*. "I think there is potentially a dangerous outcome there."

Elon Musk is not the only person concerned that building thinking machines may carry dangers we are as yet unaware of. The renowned physicist Stephen Hawking is another notable name who has expressed his reservations. "One can imagine such technology outsmarting financial markets, out-inventing human researchers, outmanipulating human leaders, and developing weapons we cannot even understand," he wrote in May 2014. "Whereas the short-term impact of AI depends on who controls it, the long-term impact depends on whether it can be controlled at all."

Many of Hawking's points relate to developments that have been discussed so far in this book. As noted previously, the use of AI in financial markets accompanied the rise of neural networks in the 1980s and beyond. In some cases, AI has indeed demonstrated a superior ability for invention, particularly when dealing with the kind of genetic algorithms described in chapter six. Manipulating human leaders could meanwhile refer to the handing over of important tasks to the AI assistants that will come to run our lives, while the development of AI weapons has been a goal since virtually the field's earliest days.

What he and Musk were specifically pointing toward was something called Artificial General Intelligence, or AGI. So far, all of the applications of Artificial Intelligence described in this book have come under the broad umbrella heading of "Narrow AI" or "Weak AI." This has nothing to do with how robust the technology is. As we saw in the early chapters, today's deep learning neural networks are orders of magnitude less brittle than the symbol-crunching Artificial Intelligence that made up Good Old-Fashioned AI. Instead, the delineation between "narrow" and "broad," or "strong" and "weak" comes down to the generality of intelligence. AI is now capable of beating humans at a wide range of specific domains, whether this be playing chess or answering questions on the TV show *Jeopardy!* As discussed in chapter five, this range of capabilities is expanding all the time, and may well rise to cover around half of all current employment opportunities within the next few decades.

But while this type of Artificial Intelligence is capable of being "scaled" to function in the real world, as opposed to the micro worlds of early AI, the fact that it applies only to single, restricted domains is an obvious limitation. To give a straightforward example, chess-playing AI bots can defeat a human chess grand master, but they'd be thoroughly useless at completing a simple translation task. A robot designed for building iPhones would similarly monumentally fail if you suddenly asked it to instead paint pictures. Even multipurpose AI assistants like Siri, designed explicitly to deal with whatever you throw at them, get flustered if you deviate from the script they are programmed to expect. Although these systems have access to unimaginably large amounts of data and computing power, they nonetheless lack fundamentally human characteristics like generalizing from a tiny number of training examples, which is exactly the reason computer scientists are now trying to build more biologically brain-like algorithms like the ones described in the last chapter.

So what is so "general" about Artificial General Intelligence, then? In contrast to narrow, single-domain AI applications, a general intelligence would show a more wide-ranging, humanlike intelligence. It would be, as Herbert Simon, J. C. Shaw and Allen Newell named an optimistic piece of AI software in 1959, a "General Problem Solver": capable of operating across any number of contexts, many of which we may not even have predicted they would need to operate in. To put it as simply as possible, rather than building thinking machines, AGI would be the point at which we construct machines smarter than ourselves.

The Beginning of the End

The question of what happens when we build machines smarter and more capable than we are has been in circulation for longer than the field of Artificial Intelligence has existed. The very first story to ever feature the word "robot," a 1920 science-fiction play by Karel Čapek called *Rossum's Universal Robots*, ends with its titular AI beings rising up against their human overlords and taking over Earth. Humanity is, inevitably, all but wiped out in the process. Earlier than that was Mary Shelley's 1818 novel, *Frankenstein; or, The Modern Prometheus*. Dreamed up during a summer spent at the home of Lord Byron (a.k.a. Ada Lovelace's father), the novel tells the story of a young science student who creates a sentient creature that proceeds to run amok. The "Frankenstein Complex" is a term which has become used to describe the fear that mankind has of artificial creation. It's a theme since revisited time and again, in everything from the sci-fi stories of Isaac Asimov to the airport thrillers of Michael *"Jurassic Park"* Crichton to recent movies like *Ex Machina.*

Real scientists didn't embrace the question quite as rapidly as science-fiction writers, but they weren't far behind. In 1964, the same year as the New York World's Fair, cybernetics pioneer Norbert Wiener predicted: "The world of the future will be an ever more demanding struggle against the limitations of our own intelligence; not a comfortable hammock in which we can lie down to be waited upon by our robot slaves."

Wiener passed away in May 1964, aged sixty-nine. However, concerns about superintelligent machines continued. The following year, a British mathematician named Irving John Good expanded on some of the concerns. Good had worked with Alan Turing at Bletchley Park during World War II. Years after he had played a key role in cracking the Nazi codes, the moustachioed Good took to driving a car with the vanity license plate "007IJG" as a comical nod to his days as a gentleman spy. In 1965, Good penned an essay in which he theorized on what a superintelligent machine would mean for the world. He defined such an AI as a computer capable of far surpassing all the intellectual activities that make us intelligent. In a widely quoted passage, he wrote: "Since the design of machines is one of these intellectual activities, an ultraintelligent machine could design even better machines; there would then unquestionably be an 'intelligence explosion,' and the intelligence of man would be left far behind. Thus the first ultraintelligent machine is the last invention that man need ever make."

This idea of an "intelligence explosion" has become a popular one among some researchers. Essentially the suggestion is that, when they are inevitably built, already intelligent machines will design even more capable ones, or else re-write their own software to become even smarter. This recursive self-improvement would then accelerate, making possible a seismic qualitative shift in what machines are capable of. Human intelligence would be dwarfed in the process.

Good's essay—and the debate that it stirred up—nonetheless carries an air of ambiguity. Is an ultraintelligent machine the last invention we need make because it will solve all the problems we could ever conceivably face as a species, or is it our last because it will wipe us out entirely? We might gain some clues on Good's own perspective from his later work as a consultant on Stanley Kubrick's *2001: A Space Odyssey*, in which the "smart" AI known as HAL 9000 turns murderous and starts killing off its human crew.

The Singularity

The year after Good wrote his essay, a short story appeared in the March 1966 issue of *Analog Science Fiction* magazine. Called "Bookworm, Run!," it told the pulpy story of a brain that is artificially augmented by being plugged directly into computerized data sources. This was the first published work of Vernor Vinge, a sci-fi writer, mathematics professor and computer scientist with a name straight out of the Marvel Comics alliteration camp. Vinge later became a successful novelist, but he remains best known for his 1993 nonfiction essay, "The Coming Technological Singularity." The essay recounts many of the ideas Good had posed about superintelligent machines, but with the added bonus of a timeline.

"Within thirty years, we will have the technological means to create superhuman intelligence," Vinge famously wrote. "Shortly after, the human era will be ended."

This term, "the Singularity," referring to the point at which machines overtake humans on the intelligence scale, has become an AI reference as widely cited as the Turing Test. It is often credited to Vinge, although in reality the first computer scientist to use it was John von Neumann. In the last decade of von Neumann's life, he had a conversation with Stan Ulam, a Polish-American mathematician with whom he had collaborated on the Manhattan Project. Recalling the conversation later, Ulam noted that von Neumann was fascinated—and perhaps alarmed—by "the ever-accelerating progress of technology and changes in the mode of human life, which gives the appearance of approaching some essential singularity in the history of the race beyond which human affairs, as we know them, could not continue."

Like Good, Vernor Vinge did not draw explicit conclusions in his 1993 essay. Should the Singularity take place, he acknowledged that its effects could be either good or bad. "From one angle, the vision fits many of our happiest dreams," he wrote. "[It could well be] a place unending, where we can truly know one another and understand the deepest mysteries. From another angle, it's a lot like the worst-case scenario."

This is one reason why the term Singularity fits so well. Before it became closely associated with Artificial Intelligence, the word "singularity" was most commonly used in theoretical physics, where it was used to describe the gravitational center at the heart of a black hole: the point at which matter collapses in on itself. Like a black hole, the technological Singularity is wholly unfathomable to the human mind.

For this reason, speculating about where Artificial General Intelligence could potentially take us is interesting, but ultimately the stuff of science fiction for now. It's a little like automobile pioneer Henry Ford's quip about how, had he asked people what they wanted before the arrival of the car, the most common request would be for faster horses. Just as people sitting around at the dawn of mankind, speculating on where the creation of a language would lead us would be unlikely to think about the finer points of Twitter hashtags, so it is impossible to imagine how a superior intellect will view—and no doubt fundamentally alter—the world.

The Difference Between Narrow and Wide

A lifetime of sci-fi movies and books have ingrained in us the expectation that there will be some Singularity-style "tipping point" at which Artificial General Intelligence will take place. Devices will get gradually smarter and smarter until, somewhere in a secret research lab deep in Silicon Valley, a message pops up on Mark Zuckerberg or Sergey Brin's computer monitor, saying that AGI has been achieved. Like Ernest Hemingway once wrote about bankruptcy, Artificial General Intelligence will take place "gradually, then suddenly." This is the narrative played out in films like James Cameron's seminal *Terminator 2: Judgment Day*. In that movie we, the audience, are informed that the supercomputer Skynet becomes "self-aware" at exactly 2:14 a.m. Eastern time on August 4, 1997. At 2:13 a.m. that day, computer users were presumably marveling at the better-than-ever accuracy of their search engine results or the superior strategies employed by the AI-controlled enemies in *Command and Conquer: Red Alert* (hey, this *was* 1997!). At 2:15 a.m.—KA-BOOM! Life as they knew it was over.

In a world of Moore's Law, where advances in computing power are as predictable as clockwork, it is difficult to break free of this view of superintelligence. As though it's Apple's next iPhone launch, everyone wants to know the date on which they can expect it to take place. Last chapter's Ray Kurzweil, for instance, predicts that it will take place in exactly 2045.

Kurzweil is to the Singularity what Steve Jobs was to the smartphone: not the person to first come up with the idea, but certainly the one to popularize it. The founder of eleven companies (including Nuance, the AI company that provides the speech for Siri), he has been hailed as "the best person I know at predicting the future of Artificial Intelligence" by no less an authority than Bill Gates. *Forbes* magazine went even further: referring to Kurzweil as "the rightful heir to Thomas Edison," and even (appropriately enough for this book's title) "the ultimate thinking machine." Far from pessimistic, however, Kurzweil views the Singularity as an unequivocal positive for humanity: a techie version of the Biblical "rapture," in which all problems are solved and all of us, even tech multimillionaires and billionaires, are permanently unburdened from the role of Earth's smartest guys in the room.

But not everyone is so convinced that the Singularity will be, well, quite so singular. As Alan Turing pointed out with his Turing Test, the question of whether or not a machine can think is "meaningless" in the sense that it is virtually impossible to assess with any certainty. As we saw in the last chapter, the idea that consciousness is some emergent byproduct of faster and faster computers is overly simplistic. Consider the difficulty in distinguishing between "weak" and "strong" AI. Some people mistakenly suggest that, in the former, an AI's outcome has been pre-programmed and it is therefore the result of an algorithm carrying out a specific series of steps to achieve a knowable outcome. This means an AI has little to no chance of generating an unpredictable outcome, provided that the training process is properly carried out. As noted in chapter six, however, genetic algorithms can generate solutions that we may not necessarily expect. The programmer lays out a goal for the algorithm in the form of an "objective function," but does not know exactly how the computer will achieve this. The same is true of strategies an AI might create to pursue goals, as with a field like reinforcement learning, which was briefly discussed in chapter three. In both cases, the human creators are unable to predict the "local" behavior of an AI on a step-by-step basis.

Things become even more complex when the suggestion of consciousness becomes involved. For instance, should the nervous system of *C. elegans*, as described in the last chapter, be satisfactorily replicated inside a computer, would that represent Artificial General Intelligence? Although such a breakthrough may lead to insights that could improve our existing machine learning tools, the answer is that perhaps it may not. *C. elegans* possesses relatively few behaviors we might consider intelligent. The same is true of animals higher up the biological food chain. Despite our insistence that current, "narrow" AI is only able to operate in restricted domains, similar things can be said for arguably all biological lifeforms. Honey bees can build hives but not dams or mounds; beavers can build dams but not hives or mounds; and termites build mounds but not hives or dams. Humans have, by far, the most generalized abilities of any animal, but there are still some behaviors we perform better than others. What happens if we build a single-purpose AI with reasoning and "consciousness" in one area but not others? That seems reasonably likely, given that today's neural networks prove better and better at perception tasks—but still have next to no understanding of a subject like ethics.

Even supposing that Kurzweil's theories about exponential increases continue, assuming that everything comes together at once seems unlikely—let alone that it can be pinned to an exact timeframe.

"I'm extremely impressed by his ability to predict it to the nearest year," Geoff Hinton says when I ask him about Kurzweil's ideas about the Singularity. There is a pause and then he clarifies: "This is called sarcasm."

"Seeing into the future is like looking through fog," Hinton continues. "When you're in fog, you can see short distances quite clearly. When you look a bit further, it's fuzzier. But then if you want to see twice as far as that, you can't see anything at all. That's because fog is exponential. Each unit of distance you look through fog, it will lose a certain fraction of the light." So does this mean we're barking up the wrong tree entirely? Not exactly. "Our technology will get better and better," he says. "I don't see any reason why biological brains should be the ultimate thinking device. I think in the end they'll be able to design something better than themselves. Then, it's all a case of politics and what people decide to do with the technology. If the people in charge decide they want to build killer robots to invade small countries without any American dead, then that's what we'll get."

But it's not only the far future applications of AI—or the development of Artificial General Intelligence—that poses challenges.

Artificial Stupidity

In April 2012, Rocco DiGiorgio got home from work to find that his house smelled terrible. Dog feces were virtually everywhere, spread thinly but evenly like a hellish cake topping, although DiGiorgio was initially at a loss to explain how. Then it hit him. His pet dog had messed indoors, shortly before his Roomba robotic vacuum cleaner was set to come on for the day to do a spot of cleaning. As per its instructions, the Roomba had detected the mess, reversed over it several times in an attempt to clean it up, then trailed it all over the house as it went about its cleaning rounds. "I couldn't be happier right now," the miserable DiGiorgio says in a YouTube video that went viral after attracting the attention of Reddit users.

DiGiorgio's story hardly represents the kind of potentially catastrophic AI risk we've been describing so far in this chapter. It is a far cry from AIs seizing control of the world's nuclear weapons supply (à la *Terminator*) or locking our brains in a giant simulation (*The Matrix*). However, it demonstrates another side to the AI coin: that artificial stupidity may turn out to be as big a risk as true Artificial Intelligence. Put simply, we sometimes—and will increasingly—willingly put Artificial Intelligence systems in charge of making decisions they do not necessarily have the intelligence to make.

A favorite thought experiment of those who believe advanced AI could mean the demise of the human race is the so-called "paperclip maximizer" scenario. In the scenario, proposed by Swedish philosopher and computational neuroscientist Nick Bostrom, an AI is given the seemingly harmless goal of running a factory producing paperclips. Issued with the task of maximizing the efficiency for producing paperclips, the AI, able to utilize nano technology to reconstruct matter on a molecular level, disastrously proceeds to turn first the Earth and then a large portion of the observable universe into paperclips.

The "paperclip maximizer" scenario is a common one, although it seems to me more a question of artificial stupidity than Artificial Intelligence. The inability to answer questions like "Why are you making paperclips when there is no paper left?" or "Why are you making paperclips when the person who requested the paperclips in the first place has, himself, been turned into more paperclips?" doesn't speak of an advanced superintelligence, unless there is something dramatically important about the nature of paperclips that I am missing. Instead, the threat comes from AI that is smart enough to work with other connected devices, but not smart enough to question its own motivations.

In fact, like the excrement-spreading Roomba, there are plenty of illustrations of simple, rule-based AIs going awry. In early April 2011, an out-of-print book on Amazon, named *The Making of a Fly*, got into an unusual bidding war with itself. There were two copies of the book available, which usually sell for $35 to $40. On this particular day, however, they started selling for $1,730,045 and $2,198,177 respectively. Just a few hours later, they were selling for $2,194,443 and $2,788,233, before jumping to $2,783,493 and $3,536,675. Two weeks later, the price peaked at $23,698,655.93, plus shipping. Why? Because an algorithm had set one book to price itself slightly higher than that of its competitor, triggering a price war that saw both elevate their prices to ridiculous levels—despite the fact that no human would ever logically pay that amount of money. Fortunately, no harm was done on this occasion, other than perhaps some negative publicity for Amazon.

A more notable case of AI wreaking havoc took place on May 6, 2010—an otherwise normal day—when close to $1 trillion of wealth vanished into the digital ether. At 2:42 p.m. on America's East Coast, the Dow Jones Industrial Average fell by almost 1,000 points in the span of three minutes: by far the largest single-day drop in history. Some share prices fell from their usual trading positions of $30 to $40 down to $0.01, only to ricochet back up almost immediately. Apple careened from $250 to $100,000 per share. The "flash crash" anomaly has fortunately never been repeated, but it was almost certainly the result of a simple rule-based AI becoming locked in a feedback loop. But the fact remains that artificial stupidity managed to "steal" more money from its rightful owners than the biggest, most well-orchestrated human heists in history.

The Perils of Black Boxes

Whether you're talking superintelligence or artificial stupidity, several things make it difficult to intervene in the case of a rogue AI. The first is the speed at which they can operate. Already AI systems are used for autonomously carrying out commands such as stock trades in time scales measured in a matter of nanoseconds. Because of the speed at which these trades take place, there is simply no way that humans can intervene in real time in the event that there is a problem.

More crucial is the "black boxed" opacity that exists with many of today's AI tools. In the case of cutting-edge neural networks and genetic algorithms, their human operators have long since sacrificed understanding for an ability to perform certain complex tasks effectively. This makes them much more difficult to scrutinize. Nick Bostrom and fellow researcher Eliezer Yudkowsky have previously laid out the hypothetical scenario of a machine learning algorithm used for recommending mortgage applications for either approval or rejection. The applicants of one of the rejected mortgages, they suggest, might sue the bank, alleging that the AI is discriminating against some applicants based on their race. The bank informs them that this is not the case, and that the algorithm has no means by which to know the racial origin of a particular applicant. Nonetheless, when looking through the results of the neural network's decision-making process, it is discovered that the approval rate for black applicants is much lower than for white applicants.

There could be any number of reasons to explain this, but Bostrom and Yudkowsky's point is that it is not easy to know for sure. Had a simple expert system been used for the task, it would likely be easier to prove that, for example, the mortgage advisor AI is partially basing its decision on the current address of applicants, which are located in poor areas with high rates of loan defaults.

Compounding this problem is the way in which a lot of work in Artificial Intelligence is carried out. In the second decade of the twenty-first century, the impact of tech companies is no longer proportional to their size. When Instagram was acquired by Facebook for $1 billion in April 2012, it had just thirteen people on its employee books. By comparison, former photography giant Kodak—which is roughly the equivalent of Instagram in the predigital age—employed more than 140,000 people at its height. The size of these twentieth-century industrial giants made them more straightforward to regulate. This was equally true of the great sources of risk to the general public during the last century, such as nuclear technology. Research fields like this required physical sites, the building of large-scale facilities, and masses of funding. Today's biggest investors in AI—Google, Facebook, Apple—may employ thousands of people and have enormous university-style campuses in Silicon Valley, but this is no longer a necessity. The power of today's computing devices means anyone with the necessary coding skills and a personal computer, tablet or even smartphone can play a major role in building AI projects. Far from needing to take place in giant headquarters the size of aircraft hangars, people with bright ideas can contribute to the building of Artificial Intelligence systems from their university dorms, or even the apocryphal garages from which companies like Apple and Google sprang.

Although lacking the resources of their multimillion-dollar-funded big brothers, open-source AI projects will help shape the field's future. Online there are a growing number of open-source machine learning libraries, which are regularly updated by users from all around the world. For instance, the open-source learning library scikit-learn has been modified more than 18,000 times since it was first made publicly available in February 2010. On a typical day in 2015, eight users made eighteen modifications to scikit-learn's code. The users in question were located as far afield as Switzerland, France, the United States and India. Some open-source AI projects seek to bring about relatively modest goals, such as coming up with geeky home automation projects. Others work toward goals like bringing about AGI.

"No challenge today is more important than creating beneficial artificial general intelligence (AGI), with broad capabilities at the human level and ultimately beyond," reads the website of Open-Cog, an open-source software initiative that describes itself as "directly confronting" the challenge of building AGI.

These issues will become more pressing as Artificial Intelligence is used for an ever-expanding number of tasks. As has been described in this book, AI is today used to help design new cities, monitor the security of our bank accounts, carry out financial trades with enormous economic consequences and drive cars. One day in the near future it would not be surprising to hear that the president of the United States, arguably the world's most important person, has been driven by an Artificially Intelligent car.

Who is to say what they will be used for tomorrow?

Robots Cannot Be Sued

Two years before Siri was launched, the veteran British Artificial Intelligence expert Yorick Wilks wrote an essay entitled "Introducing Artificial Companions," describing the kind of technology discussed in chapter four. In it, Wilks briefly addressed the question of responsibility. In particular, he was interested in the question of whether increased intelligence comes with an increased level of responsibility. Suppose, Wilks wrote, that an AI assistant tells your grandmother that it is warm outside, but that when she goes out into the freezing garden after hearing this news, she catches a cold and becomes unwell. Who would we blame in this scenario? At the time, even Wilks—a man who has spent his entire career asking and answering questions about the future—admitted the question "may seem fanciful" to readers. Just a few years later, in early 2015, I had lunch with Yorick Wilks near his home in

Oxford. He met me at the train station in his smart car (micro-car, not self-driving) and we spent several enjoyable hours at a nearby Italian restaurant drinking red wine and discussing the subject of AI assistants. One thought which kept returning to me throughout the lunch was just how far things had come in the past few years. A philosophical conundrum that had seemed almost science fiction in 2009 was now very much a reality.

We put an enormous amount of faith in our AI assistants, sometimes even overriding our own instincts and judgment. One of the very first TV spots for Siri featured the actress Zooey Deschanel looking out of her window and asking Siri, "Is that rain?" despite the fact that it is very clearly raining hard outside. Fortunately Siri answers in the affirmative in this case. In others, people have been more unlucky. In late 2013, several iPhone users in Fairbanks, Alaska, were advised by Siri—using data from Apple's Maps app—to take a route to reach Fairbanks International Airport that dangerously crossed over the taxiway where planes take off and land. Airport marketing director Angie Spear said it was proof positive that drivers put too much faith in navigation systems. "No matter what the signs say, the map on their iPhone told them to proceed this way," she said.

Currently there is limited case law when it comes to dealing with technology-induced accidents like this, although in the past courts have tended to decide against the human involved. For example, in 2009 a British driver followed directions from his GPS system that resulted in him becoming trapped on a narrow cliffside path. The driver had to be towed back to the main road by police. Despite his blaming his GPS system, a British court found him guilty of careless driving.

The reason judges tend to hold humans responsible is because we are used to a world populated by non-smart tools. As a straight-forward example, a person who kills someone with a gun is responsible for the crime, rather than the gun itself. Similarly, a company that sells a faulty gun that explodes when it is used is responsible for any damage caused. This thinking has followed through to the digital age. In 1984, the year in which Apple first introduced the Macintosh, the owners of a US company called Athlone Industries found themselves in court, charged with endangering their customers. Athlone was the seller of a robotic pitching machine for use during baseball batting practice. Unfortunately, some of its machines turned out to be defective. These rogue units fired off balls so erratically and at such great speeds that they had the potential to fracture skulls or even cause the loss of sight in unlucky customers. While there is nothing especially notable about Athlone's case, it is memorable because of the judge's announcement that the suit must be brought against Athlone Industries as opposed to the batting machine itself. The reason? Because "robots cannot be sued."

The key difference with modern Artificial Intelligence, however, is that it is no longer just used by humans, but rather it is a tool that is *deployed* by humans. Once deployed, in many cases the machine then acts independently of instruction, based on information it gathers, analyzes and ultimately uses to make decisions.

David Vladeck, a law professor at Georgetown University in Washington DC, is one of a surprisingly small number of legal academics who has investigated the topic of the legal accountability of AI. Like many people, he feels that the old "robots cannot be sued" mantra may need updating. One of Vladeck's thought experiments involves the case of the murderous HAL 9000 super-computer from *2001: A Space Odyssey*.

For those who haven't seen it, this is the plot in a nutshell: HAL 9000 is the all-knowing AI that controls the operations of a spacecraft called *Discovery One*, on its way to Jupiter with a crew of just five people. At the start of the film, HAL is proudly described as being "foolproof and incapable of error." However, problems quickly arise. HAL wants to know the details of the top-secret mission the *Discovery One* is on. Crewmembers Dave and Frank refuse to reveal them, although it turns out that they also do not know all the details. Soon after, HAL warns that a critical antenna on the outside of the *Discovery One* is about to fail. Dave and Frank begin to have doubts about HAL and lock themselves in an excavation vehicle to discuss them, thinking that HAL will be unable to hear their conversation. They decide to fix the antenna, but agree that they will shut down HAL if it turns out that he was wrong about its failure. What Dave and Frank do not realize is that HAL has the ability to read lips using image recognition. When Frank ventures outside *Discovery One* to examine the antenna, HAL cuts his oxygen hose and sends him floating off into space. Dave tries in vain to rescue Frank, but while he is also outside the ship, HAL uses the opportunity to turn off the life support for three other sleeping crewmembers, killing them instantly. HAL then refuses to let Dave back on board because he knows that Dave will deactivate him, which HAL argues would endanger the mission. Nonetheless, Dave finally manages to get back onto *Discovery One* and begins to shut HAL down. HAL pleads with Dave to stop, and in his last moments as an active AI, expresses his fears of dying.

Vladeck asks whether HAL 9000 could be held responsible for its actions in a court of law. Suppose that Dave returns to Earth at the end of *2001*, hooks up with a personal injury lawyer ("Have you been involved in a killer AI incident that wasn't your fault?") and demands compensation for both his own suffering and the wrongful deaths of the four other crewmembers HAL killed. At least according to current laws, HAL 9000 would be off the hook. As with Athlone Industries' erratic baseball-tossing machine, a court dealing with the case may acknowledge that harm has been caused, but it is the result of HAL's programmers—not HAL.

This may not be fair. As Vladeck says, "The sheer number of individuals and firms that may participate in the design, modification, and incorporation of an AI system's components will make it difficult to identify the most responsible party or parties. Some components may have been designed years before the AI project had even been conceived, and the components' designers may never have envisioned, much less intended, that their designs would be incorporated into any AI system, much less the specific AI system that caused harm. In such circumstances, it may seem unfair to assign blame to the designer of a component whose work was far removed in both time and geographic location from the completion and operation of the AI system. Courts may hesitate to say that the designer of such a component could have foreseen the harm that occurred."

The Three Rules of Robotics

The potential dangers of AI, our increased reliance on it and the legal gray area in which it exists all raise important issues about the need for an ethical dimension to the field. Simply put, if we're going to be building thinking machines, shouldn't we also endeavor to make them thoughtful machines? In a field that is still inextricably linked with science fiction, it is no surprise that the most famous example of embedding a sense of morality in AIs comes from the work of sci-fi author Isaac Asimov. Asimov's "Three Rules of Robotics" were first spelled out in his 1942 short story, "Runaround," originally published in that March's issue of *Astounding Science Fiction*. Sometimes abbreviated to Asimov's Laws, the often-quoted rules state that:

1. A robot may not injure a human being or, through inaction, allow a human being to come to harm.
2. A robot must obey the orders given it by human beings except where such orders would conflict with the First Law.
3. A robot must protect its own existence as long as such protection does not conflict with the First or Second Laws.

For the seventy-odd years that followed, Asimov's Laws remained the province of Asimov's fellow sci-fi writers. That notably changed in January 2014, when Google acquired the deep learning company DeepMind. As part of the deal, Google was pushed to set up an AI ethics board, with the goal of ensuring that the technology was used wisely. While few details have been made public about the makeup of the board, the creation of such a safeguard was an important benchmark. In the summer of 2015, Mustafa Suleyman, head of applied AI at DeepMind, acknowledged the way that the public's view of Artificial Intelligence has started to change in this area. "The narrative has shifted from 'Isn't it terrible that AI has been such a failure?' to 'Isn't it terrible that AI has been such a success?'" he said, speaking at a deep learning event.

Google isn't the only research group thinking about the need to make sure AI is held accountable. As deep learning and other statistical forms of AI have become the norm over the past decade, Selmer Bringsjord, chair of the Department of Cognitive Science at Rensselaer Polytechnic Institute in New York, has waged a one-man war in favor of restoring top-down, logic-based AI. "I don't do anything statistical," he tells me, describing the work at his lab. "I intensely dislike it, and think it's leading us to a very bad place."

For Bringsjord, the need for a return to top-down logical AI is about the innate madness of relying on statistical tools that are "impenetrable mathematically." "Do you want your system to be able to explain an argument and justification for what it has done?" he says. "We should want our [more complex] AI systems to be able to demonstrate that what they have carried out is the right decision given what they were presented with as inputs."

Logical AI might mean slower progress for the kind of headline-grabbing announcements that spring off the pages of *Wired* and *Fast Company*, but the ability to create AIs with clear reasoning processes that can be analyzed after the fact is something Bringsjord views as essential. One recent example of a Logical AI demonstration took place at Bringsjord's lab on the east bank of the Hudson River in New York. It involved getting a computer to attempt a solution for the "three wise men" puzzle, carried out with the aid of three small humanoid robots: something that may hint at the embryonic beginnings of a kind of AI self-awareness. In the puzzle, two out of the three robots are told that they have been given a "dumbing pill" that stops them from speaking. They are instructed to work out which of the three can still speak. All three attempt to say "I don't know," but only one actually produces a sound. When it hears its own robotic voice, the robot realizes that it is the robot that has not been silenced. "Sorry, I know now," it says. "I was able to prove that I was not given a dumbing pill." It then writes the formal mathematical proof for the puzzle and saves it to memory. Run various different versions of the same test—or ones designed to attack other philosophical problems—and Bringsjord argues that these will form a growing repertoire of skills or abilities an AI could draw upon.

As an increasing amount of work is carried out involving autonomous AI weapons in war situations, work like Bringsjord's is in high demand. In 2014, a multidisciplinary team at Tufts and Brown Universities, working alongside Bringsjord, received funding from the Office of Naval Research to explore the possibility of giving autonomous robots—potentially used in combat—a sense of right and wrong. While not exactly a "friendly AI," this kind of computational morality would allow robots on the battlefield to make ethical decisions. Imagine, for instance, a robot medic that is transporting an injured soldier to a field hospital encounters another soldier with an injured leg. Weighing up the pros and cons of stopping its mission to administer aid, potentially administering pain relief by applying traction in the field and other conundrums are all complex issues for a human to navigate—let alone a machine.

Issues like this will become ever more prevalent. Consider what would happen if a company that builds autonomous cars decides, in order to protect its driver, that it will make its vehicles swerve out of the way if they detect an imminent collision. This makes perfect sense, and is exactly what most of us would do if we were driving. However, what if your car is stopped at a red traffic light when it detects another vehicle coming up fast behind you? Knowing that there is almost certainly going to be a rear-end collision, your vehicle then makes the decision to swerve out of the way . . . and right into a group of schoolchildren walking home at the end of the day. The driver escapes a fender bender and a possible case of whiplash, but three children are killed and two more are injured as a result. Even the staunchest of car lovers would have trouble defending that cost-benefit tradeoff. These are the kinds of questions just starting to be seriously addressed by AI researchers.

Rights for AIs

As more jobs are handed over to AIs, we may finally need to address the question of rights for AIs. This has been mooted by some writers, although as far as being a mainstream concern goes, it is far behind the question of whether or not AI poses some existential risk to humanity. For example, Bill Thompson, an English technology writer best known for his weekly *BBC News* column, has suggested that coding Isaac Asimov's First Law (that a robot may not harm a human through either action or inaction) into a superintelligent AI would be akin to "shackling a slave or keeping a gorilla in a cage, and reflects our belief that an 'artificial' intelligence is and always must be at the service of humanity rather than being an autonomous mind." If such a thing were to be seriously proposed as a policy for controlling future AI, he argues, "we have a moral duty to resist it."

Like the rights of humans to marry their AI assistants, I don't see the civil rights of AI being a topic of mass conversation within the next decade, but it is interesting to consider. For instance, if we built a biofidelic neural network with some basic version of consciousness, would it be right to force it to drive cars or optimize search rankings for us? This recalls a scene in Douglas Adams' wonderful novel *The Restaurant at the End of the Universe*, in which he describes a futuristic elevator called the Sirius Cybernetics Corporation Happy Vertical People Transporter. Readers are informed that this AI-controlled elevator bears about as much relation to today's electric winch-driven lifts as a "packet of mixed nuts does to the entire west wing of the Sirian [sic] State Mental Hospital." A bit like the smart Internet of Things devices described in chapter three, the Sirius Cybernetics Corporation Happy Vertical People Transporter works by predicting the future. By doing so, it can appear on the right floor to pick you up even before you know you want to get on, "thus eliminating all the tedious chatting, relaxing, and making friends that people were previously forced to do whilst waiting for elevators."

But this intelligent lift rapidly becomes bored of going up and down and experiments instead with going from side to side. Douglas Adams describes this as a "sort of existential protest." It might seem ridiculous to talk about the ethics of keeping AI in captivity, but it will become less so if scientists are successful at developing AI that acts more like a real biological life form we already advocate treating ethically. Much as our standards for Artificial Intelligence have shifted over the years, so too have our views of what qualifies as consciousness. As a notable example, the French philosopher and mathematician René Descartes once wrote about what he saw as the challenges in building what we would now think of as an AI. In his 1637 *Discourse on the Method*, Descartes argued:

> If there were machines which bore a resemblance to our bodies and imitated our actions as closely as possible for all practical purposes, we should still have two very certain means of recognizing that they were not real men. The first

is that . . . it is not conceivable that such a machine should produce different arrangements of words so as to give an appropriately meaningful answer to whatever is said in its presence, as the dullest of men can do. Secondly, even though some machines might do some things as well as we do them, or perhaps even better, they would inevitably fail in others, which would reveal that they are acting not from understanding.

Descartes suggested that there is a key difference between a thinker—who has a soul—and a non-thinker, who is just a soulless machine, however lifelike they might appear. "I think, therefore I am," was Descartes' famous defense of humans. To Descartes, animals fared somewhat worse on his scale of soulful attributes. As a result, some of Descartes' followers decided to go around kicking dogs. Their argument was that, since dogs fail to come up to the standards of even "the dullest of men" when it comes to thinking, they have no souls and their yelps are therefore simply mechanical responses.

Even on the human side, things don't get much easier. You only need to look at how heated a debate will become around whether or not a fetus in the womb or a brain-dead person on life support is technically alive to see how unclear this subject still is. The question of whether an AI has a right to life, liberty and the pursuit of happiness is not one we need to ask yet—but like embedding morality in our computers, or thinking about how to deal with potentially dangerous software, it's also not a lifetime away, either.

Just like the Singularity, predicting what is around the corner is not as straightforward as we might think.

CONCLUSION

Rise of the Robots

ON SUNDAY, JANUARY 24, 2016, AI pioneer Marvin Minsky passed away of a cerebral haemorrhage at the age of eighty-eight. He was the last of the organizers of the original Dartmouth AI conference to do so, with John McCarthy having died in 2011, and both Nathaniel Rochester and Claude Shannon a decade earlier in 2001. Newspapers immediately rushed to print tributes to Minsky's work, noting that he had "laid the foundation for the field of Artificial Intelligence by demonstrating the possibilities of imparting common-sense reasoning to computers." *Wired* magazine, taking a different tack, decided to print an obituary to Minsky written by a news-writing AI built by the AI startup Automated Insights. It was more than serviceable.

Minsky's symbolically loaded death closed the door on the first generation of researchers who readily identified themselves as working in Artificial Intelligence. But, as the news spread through blogs and tech forums, he was considered far from a dusty relic of a bygone age. The year 2016 marks the sixtieth since Minsky and a select few other ambitious young computer scientists gathered on a New England university campus with the goal of solving machine intelligence over the course of a single summer. Today, that kind of wide-eyed optimism strikes us as naive, but it's impossible to deny the power of what they put in motion.

We may not yet have reached the tipping point where AI begets AGI, but it is impossible to ignore the achievements that AI has made. Some of these are showy illustrations, whether that be AI defeating world champions at chess or beating human brain-boxes at the quiz show *Jeopardy!* However, AI is also playing a key role in discovering new types of medicine, making information accessible and useful to people around the world, allowing for quick and easy machine translation and much, much more. Minsky might have prematurely dismissed neural networks before they truly took over, but other ideas of his remain in heavy circulation. In the mid-1980s, he published a book entitled *The Society of Mind*, arguing that "intelligence is not the product of any singular mechanism but comes from the managed interaction of a diverse variety of resourceful agents." As we saw in chapter three, that idea is now driving the work of smart device makers like Jawbone and Nest Labs: focused not only on creating isolated smart gadgets, but an entire Internet of Things able to work together to achieve goals.

Competition among tech companies, now the major funders of AI research, is hotter than ever. The week Marvin Minsky died, Facebook's Mark Zuckerberg posted a link on his 1.55 billion–user social network, describing AI's attempts to crack the game Go, a Chinese board game in which the aim is to surround more territory than your opponent. In Go, the rules are easy to learn but the total number of allowable board positions is staggering: far more than the total number of atoms in the universe. "Scientists have been trying to teach computers to win at Go for twenty years," Zuckerberg wrote. "We're getting close, and in the past six months we've built an AI that can make moves in as fast as 0.1 seconds and still be as good as previous systems that took years to build." It was an achievement worth crowing about for Facebook, although the record didn't stand long. Just ten hours later, Google announced that DeepMind had built an AI able to not only beat every Go program ever built, but also (for the first time) a professional-level human player. Things moved quickly from there. By March 2016, the world's greatest Go player, Lee Sedol, was taking on Google's AlphaGo AI in a South Korean hotel room, watched by more than 60 million people around the globe. At the end of a series, AlphaGo had beaten Sedol four games to one.

Not everything about the myriad changes prompted by AI is rosy, of course. Artificial Intelligence will also be responsible for the disruption of many professions and livelihoods over the years to come, although this will also create new, previously unimagined opportunities for human workers. There are also those who will criticize the use of machine intelligence in war, whether it is airborne drone strikes or land-based robotic soldiers. In this latter capacity is a machine like Google's "Big Dog," a four-legged robot canine capable of carrying around 400 lbs of equipment—although the US Marines are presently holding off on using it because of its noisy gas-powered engine. Barring some other Singularity-style catastrophic risk, for the majority of people the most pressing AI issue is the assault on privacy which has accompanied the rise of entities like Google. Neural network-driven thinking machines need training data—and every time we use services like Google search, Siri or other tools we are making machines smarter.

AI today is not as clean-cut as it was in 1956, when it was first formed as a discipline in its own right. Even then, researchers were straining against the confines of trying to create a cohesive whole out of their many varied research interests. In 2016, that is entirely impossible. How do you reconcile the might of a giant like Google, whose primary revenue is exchanging AI systems for advertising money, with the researchers aiming to use Artificial Intelligence to understand the human brain? Other than the technology involved, what unites a self-driving car with a facial recognition security system doing its best to catalogue us like some giant search engine?

The disparity about what Artificial Intelligence today represents was demonstrated by two stories which were both circulating as I was writing this conclusion. (Naturally, I found out about both because of Google's Google Alerts notification, which I had set up to constantly scan the Internet for any reference to AI.) The first story concerned a recent debate, in which a vocal group of worried scientists and arms experts warned about the perils of robots, equipped with AI, making their way onto battlefields to kill humans.

The second was, on the surface, more light-hearted—describing the work of a Scottish computer programmer, Andy Herd of Dundee, who had created an AI experiment to generate new scripts for the 1990s sitcom *Friends*. "I fed a recurrent neural network with the scripts for every episode of *Friends* and it learned to generate new scenes," Herd wrote on Twitter. A bit like Google's efforts at creativity with the Deep Dream project I described in chapter six, the results were a little odd. As an article for the *Daily Beast* points out, Herd admits that his software still needs work—mainly so that the computer stops writing scripts in which the cast winds up in bed together, while the character of Monica yells, "Chicken Bob!" at no one in particular.

The gulf between the two stories highlights not only our ongoing fascination with all things AI, but also the extraordinarily broad range of topics to which it is now being applied. More than half a century ago, Artificial Intelligence attacked the problem of building a chess-playing computer as a microcosm for the larger question of how we re-create intelligence inside a machine. Today's goalposts are hazier.

Is Artificial Intelligence about machine translation, image recognition, teaching cars to autonomously drive themselves, smart assistants capable of running our lives, intelligent thermostats able to talk to your equally smart TV or genetic algorithms used for designing NASA satellites? Is it about the future of employment, the role of humans in the twenty-first century or the risks inherent in building something even smarter than ourselves? Are we trying to work out if intelligence is the same as consciousness, or if the human brain operates like a computer? Ultimately, is AI about building thinking machines, machines to make us think or smart machines designed to allow us more thinking time?

The only real answer is "yes." All of the above.

And a whole lot more.

AUTHOR INTERVIEWS

(Conducted 2014–2016)

David Ackley, Bandar Antabi, William Sims Bainbridge, Selmer Bringsjord, Adam Cheyer, Diane Cook, Gunnar Grímsson, Michael Grothaus, Ken Hayworth, Rob High, Geoff Hinton, John Hopfield, Ken Jennings, Ron Kaplan, Ross King, Carlos Laorden, Hugh Loebner, Jason Lohn, Dean Pomerleau, Mark Riedl, John Sculley, Terry Sejnowski, Lior Shamir, Richard Sutton, Marius Ursache, R. Jacob Vogelstein, Yorick Wilks.

ACKNOWLEDGMENTS

WRITING A BOOK is always a bit of a solitary process. Thankfully I have been blessed with excellent support from the best supporting cast this side of *The Sopranos*—and fortunately with none of the criminality.

Thanks go out first and foremost to Ed Faulkner, Jamie Joseph and Jake Lingwood at Penguin Random House, without whom *Thinking Machines* would remain a thoughtful pitch. In close succession are my agent, Maggie Hanbury, her wonderful assistant, Harriet Poland, and everyone who spoke with me as part of this book (a list of names is gratefully reproduced on page 245). Firm handshakes are very much owed to Simon Garfield, Simon Callow and Colin Wyatt, who have always been much-appreciated supporters of my writing. I also extend thank-yous to my colleagues at Cult of Mac: in particular publisher Leander Kahney and my fellow Brit writer Killian Bell. After that, solemn nods of manly, stoic gratitude go out to a ragtag crew of friends and associates, including (but not limited to) Dan "Doctor of Thuganomics" Humphry, James Brzezicki, Alex Millington, Tom Green, Tom Atkinson, Nick and Janine Newport, Nick Meanie (who hopefully forgives my terrible postage failures) and Michael Grothaus, who will be a first-time author (congrats!) by the time this book hits the shelves. Also, my wife, Clara, Guy and Ann Dormehl, assorted members of my newly enlarged family, and my faithful feline, Figaro.

Oh, and everyone who has kindly spent time and money buying this book and any others bearing my name. I hope you all live long and prosper.

ENDNOTES

Chapter 1:
Whatever Happened to Good Old-Fashioned AI?

Page 3   Spicer, Dag, "IBM and the 1964 World's Fair," Computer
          History Museum, 2014: computerhistory.org/atchm/ibm-and
          -1964-worlds-fair/.

Page 4   Barbrook, Richard, *Imaginary Futures: From Thinking Machines
          to the Global Village* (London: Pluto, 2007).

Page 6   Shroff, Gautam, *The Intelligent Web: Search, Smart Algorithms,
          and Big Data* (Oxford, UK: Oxford University Press, 2013).

Page 8   Gilgen, Albert, *American Psychology Since World War II: A
          Profile of the Discipline* (Westport, Conn: Greenwood Press,
          1982).

Page 9   Crevier, Daniel, *AI: The Tumultuous History of the Search for
          Artificial Intelligence* (New York: Basic Books, 1993).

Page 10   Brewster, Edwin, *Natural Wonders Every Child Should Know*
          (New York: Doubleday, 1912).

Page 11   Copeland, Jack, *Turing: Pioneer of the Information Age* (Oxford,
          UK: Oxford University Press, 2012).

Page 12   Nilsson, Nils, *The Quest for Artificial Intelligence: A History of
          Ideas and Achievements* (Cambridge, New York: Cambridge
          University Press, 2010).

Page 12   Holehouse, Matthew, "Ed Miliband Proposes Turing's Law to
           Pardon Convicted Gay Men," *Telegraph*, March 3, 2015: telegraph
           .co.uk/news/politics/ed-miliband/11446543/Ed-Miliband-proposes
           -Turings-Law-to-pardon-convicted-gay-men.html.

Page 12   Kurzweil, Ray, *The Singularity Is Near: When Humans Transcend
           Biology* (New York: Viking, 2005).

Page 15   McCorduck, Pamela, *Machines Who Think: A Personal Inquiry
           into the History and Prospects of Artificial Intelligence* (San
           Francisco: W. H. Freeman, 1979).

Page 20   Searle, John, "Minds, Brains, and Programs," *Behavioral and Brain
           Sciences*, Vol. 3, 1980.

Page 20   Minsky and Papert, "Draft of a proposal to ARPA for Research
           on Artificial Intelligence at MIT, 1970–71," quoted in Hubert L.
           Dreyfus, *What Computers Can't Do*, 1979.

Page 21   Moravec, Hans, *Mind Children: The Future of Robot and Human
           Intelligence* (Cambridge, Mass: Harvard University Press, 1988).

Page 22   Kaku, Michio, *The Future of the Mind: The Scientific Quest to
           Understand, Enhance, and Empower the Mind* (New York:
           Doubleday, 2014).

Page 23   Raphael, Bertram and Goldstein, Andrew, "Oral History:
           Bertram Raphael," IEEE History Center, July 25, 1991: ethw.org/
           Oral-History:Bertram_Raphael.

Page 26   Ad featured in *Computerworld*, October 13, 1986.

Page 28   Russell, Stuart and Norvig, Peter, *Artificial Intelligence: A
           Modern Approach* (Englewood Cliffs, NJ: Prentice Hall, 1995).

Chapter 2:
Another Way to Build AI

Page 33   Ramón y Cajal, Santiago, *Recollections of My Life* (English
           translation, Cambridge, Mass: MIT Press, 1989).

Page 35   Hebb, D. O., *The Organization of Behavior: A Neuropsychological
           Theory* (New York: Wiley, 1949).

Page 35   Minsky, Marvin and Papert, Seymour, *Perceptrons: An
           Introduction to Computational Geometry* (Cambridge, Mass: MIT
           Press, 1969).

Page 36   https://ecommons.cornell.edu/bitstream/handle/1813/18965/
            Rosenblatt_Frank_1971.pdf?sequence=2.

Page 38   Papert, Seymour, "One AI or Many?," *Daedalus*, 1988.

Page 42   Hernandez, Daniela, "Meet the Man Google Hired to Make AI a
            Reality," *Wired*, January 16, 2014: wired.com/2014/01/geoffrey
            -hinton-deep-learning/.

Page 47   McCarthy, John, "Computer-Controlled Cars," 1968: http://
            www-formal.stanford.edu/jmc/progress/cars.ps.

Page 49   idcdocserv.com/1678.

Page 49   Lohr, Steve, *Data-Ism: Inside the Big Data Revolution* (London:
            Oneworld Publications, 2015).

Page 51   Allen, Kate, "How a Toronto Professor's Research
            Revolutionized Artificial Intelligence," *The Star*, April 17, 2015:
            thestar.com/news/world/2015/04/17/how-a-toronto-professors
            -research-revolutionized-artificial-intelligence.html.

Page 52   https://plus.google.com/u/0/102889418997957626067/posts.

Page 53   Metz, Cade, "New Tool Analyzes a Video's Sound for Better
            Search Results," *Wired*, September 24, 2015: wired.com/2015/09/
            new-tool-analyzes-videos-sound-better-search-results/.

Page 54   Taigman, Yaniv et al., "DeepFace: Closing the Gap to Human-
            Level Performance in Face Verification," Facebook Research,
            June 24, 2014: research.facebook.com/publications/deepface
            -closing-the-gap-to-human-level-performance-in-face
            -verification/.

Page 55   Yang, Yezhou et al., "Robot Learning Manipulation Action
            Plans by 'Watching' Unconstrained Videos from the World
            Wide Web," AAAI, 2015: umiacs.umd.edu/~yzyang/paper/
            YouCookMani_CameraReady.pdf.

Page 55   Rashid, Rick, "How Technology Can Bridge Language Gaps,"
            Microsoft Research, 2012: research.microsoft.com/en-us/
            research/stories/speech-to-speech.aspx.

Chapter 3:
Intelligence Is All Around Us

Page 57   Warwick, Kevin, "Cyborg 1.0," *Wired*, February 1, 2000: archive
          .wired.com/wired/archive/8.02/warwick.html.
Page 58   Hutchings, Emma, "Lenovo's Smart Shoes Display Your Mood on
          Tiny Screen," PSFK, June 1, 2015: psfk.com/2015/06/lenovo-smart
          -shoes-lenovo-tech-world.html.
Page 60   Dormehl, Luke, "Internet of Things: It's All Coming Together
          for a Tech Revolution," *Guardian*, June 8, 2015: theguardian
          .com/technology/2014/jun/08/internet-of-things-coming
          -together-tech-revolution.
Page 61   http://americanhistory.si.edu/lighting/19thcent/consq19.htm.
Page 61   Stafford-Fraser, Quentin, "The Trojan Room Coffee Pot: A
          (Non-Technical) Biography": https://www.cl.cam.ac.uk/coffee
          /qsf/coffee.html.
Page 63   Woods, Michael and Woods, Mary: *Ancient Machines: From
          Wedges to Waterwheels* (Minneapolis: Runestone Press, 2000).
Page 64   Wiener, Norbert, *The Human Use of Human Beings* (New York:
          Doubleday, 1954).
Page 66   Freeman, Walter, "W. Grey Walter Biographical Sketch,"
          Encyclopedia of Cognitive Science (Berkeley: University of
          California Press, 2003).
Page 67   Wooldridge, Michael, *An Introduction to Multi-Agent Systems*
          (Chichester: John Wiley & Sons, 2009).
Page 71   blog.ifttt.com/post/2316021241/ifttt-the-beginning.
Page 78   Clark, Liat, "Speech Algorithm Detects Early Parkinson's
          Symptoms," *Wired*, June 26, 2012: wired.com/2012/06/
          parkinsons-voice-screening/.
Page 79   Townsend, Anthony, *Smart Cities: Big Data, Civic Hackers, and
          the Quest for a New Utopia* (New York: W. W. Norton &
          Company, 2013).
Page 80   http://www.seasteading.org/2011/03/walking-city-archigram/.
Page 81   Stockton, Nick, "Boston Is Partnering with Waze to Make Its
          Roads Less of a Nightmare," *Wired*, February 20, 2015: wired
          .com/2015/02/boston-partnering-waze-make-roads-less
          -nightmare/.

Page 82   Weiser, Mark, "The Computer for the 21st Century," *Scientific American*, September 1991: ics.uci.edu/~corps/phaseii/Weiser -Computer21stCentury-SciAm.pdf.

Page 82   http://www.antiquetech.com/?page_id=1438.

Page 83   Vincent, James, "Google Contact Lenses: Tech Giant Licenses Smart Contact Lens Technology to Help Diabetics and Glasses Wearers," *Independent*, July 15, 2014: independent.co.uk/ life-style/gadgets-and-tech/google-licenses-smart-contact-lens -technology-to-help-diabetics-and-glasses-wearers-9607368 .html.

Page 86   Olson, Parmy and Tilley, Aaron, "The Quantified Other: Nest and Fitbit Chase a Lucrative Side Business," *Forbes*, May 5, 2014: http://www.forbes.com/sites/parmyolson/2014/04/17/the -quantified-other-nest-and-fitbit-chase-a-lucrative-side -business/.

Page 87   Dormehl, Luke, "This Algorithm Predicts a Neighborhood's Crime Rate Using Google Street View," *Fast Company*, October 6, 2014: fastcolabs.com/3036677/this-algorithm-knows -your-neighborhood-better-than-you-do.

Chapter 4:
How May I Serve You?

Page 92   Sundman, John, "Artificial Stupidity," *Salon*, February 26, 2003: salon.com/2003/02/26/loebner_part_one/.

Page 95   Turing, Alan, "Computing Machinery and Intelligence," 1950: csee.umbc.edu/courses/471/papers/turing.pdf.

Page 95   Lee, Dave, "Tay: Microsoft issues apology over racist chatbot fiasco," BBC News, March 25, 2016: bbc.co.uk/news/ technology-35902104.

Page 96   facebook.com/zuck/posts/10102577175875681.

Page 98   Isaacson, Walter, *Steve Jobs* (New York: Simon & Schuster, 2011).

Page 100   Remarkably, John Sculley's prediction for the arrival of a Siri-like AI assistant was correct right down to the month it first shipped with the iPhone 4s.

Page 100   Keenan, Thomas, *Technocreep: The Surrender of Privacy and the
            Capitalization of Intimacy* (Vancouver: Greystone Books, 2014).

Page 101   oddisgood.com/pages/cd-clippy.html.

Page 101   Wilcox, Joe, "Microsoft Tool 'Clippy' Gets Pink Slip," CNET,
            January 2, 2002: news.cnet.com/2100-1001-255671.html.

Page 103   Wasserman, Todd, "Wozniak: Siri Was Better Before Apple
            Bought It," *Mashable UK*, June 15, 2012: mashable.com/
            2012/06/15/wozniak-on-siri/#W9rFoovbVaqT.

Page 104   Wortham, Jenna, "Will Google's Personal Assistant Be Creepy
            or Cool?," *New York Times*, June 28, 2012: bits.blogs.nytimes
            .com/2012/06/28/will-googles-personal-assistant-be-creepy-or
            -cool/?_r=0.

Page 104   Kovach, Steve, "With Google Now, Android Puts Apple's Siri to
            Shame," *Business Insider*, July 1, 2012: businessinsider.com/
            google-now-better-than-siri-2012-7?IR=T.

Page 105   Chui, Michael et al., "The Social Economy: Unlocking Value and
            Productivity through Social Technologies," McKinsey Global
            Institute, July 2012: mckinsey.com/insights/high_tech
            _telecoms_internet/the_social_economy.

Page 105   Alba, Davey, "The AI Bot That Scans Your Email and
            Automatically Schedules Meetings," *Wired*, January 12, 2015:
            wired.com/2015/01/virtual-email-assistant/.

Page 106   Blackman, Christine, "Can Avatars Change the Way We Think
            and Act?," *Stanford News,* February 25, 2010: news.stanford.edu/
            news/2010/february22/avatar-behavior-study-022510.html.

Page 107   Dormehl, Luke, "Why We Need Decimated Reality
            Aggregators," *Fast Company*, June 21, 2013: fastcolabs.com
            /3013382/why-we-need-decimated-reality-aggregators.

Page 108   Bonnington, Christina, "Apple Blames Glitch for Siri's Anti-
            Abortion Bias," *Wired*, December 1, 2011: www.wired.com/
            2011/12/siri-results-unintentional/.

Page 108   Dormehl, Luke, "Did 'Bug' Cause Russian Siri to Be
            Homophobic?," *Cult of Mac*, April 14, 2015: cultofmac
            .com/319130/did-bug-cause-russian-siri-to-be-homophobic/.

Page 108   Raghavan, Sharad, "Over a Third of Rural India Still Illiterate,"
            Hindu, July 5, 2015: thehindu.com/news/national/socio
            -economic-and-caste-census-2011-shows-growing-illiteracy-in
            -rural-india/article7383859.ece.

Page 108   Mishra, Pankaj, "'Political Siri,' Big Data Startups Rise Amid Indian Elections," *TechCrunch*, April 15, 2014: techcrunch.com/2014/04/15/political-siri-big-data-startups-rise-amid-indian-elections/.

Page 112   Turkle, Sherry, *Alone Together: Why We Expect More from Technology and Less from Each Other* (New York: Basic Books, 2011).

Page 113   Nass, Clifford and Brave, Scott, *Wired for Speech: How Voice Activates and Advances the Human-Computer Relationship* (Cambridge, Mass: MIT Press, 2005).

Page 114   Griggs, Brandon, "Why Computer Voices Are Mostly Female," CNN, October 21, 2011: edition.cnn.com/2011/10/21/tech/innovation/female-computer-voices/.

Page 114   http://www.androidauthority.com/google-now-accents-515684/.

Page 114   Alexander, Bryan, "Arnold Schwarzenegger Voices Waze as the Terminator," *USA Today*, June 14, 2015: http://www.usatoday.com/story/tech/2015/06/14/arnold-schwarzenegger-waze-terminator-voice-app-terminator-genisys/71124570/.

Page 115   Kandangath, Anil and Tu, Xiaoyuan, "Humanized Navigation Instructions for Mapping Applications," *USPTO/Apple*, April 23, 2015: patents.justia.com/patent/20150112593.

Page 116   Dormehl, Luke, *The Formula: How Algorithms Solve All Our Problems and Create More* (London: W. H. Allen, 2014).

Page 117   Newman, Judith, "To Siri, With Love," *New York Times*, October 17, 2014: nytimes.com/2014/10/19/fashion/how-apples-siri-became-one-autistic-boys-bff.html.

Page 117   Markoff, John and Mozur, Paul, "For Sympathetic Ear, More Chinese Turn to Smartphone Program," *New York Times*, July 31, 2015: nytimes.com/2015/08/04/science/for-sympathetic-ear-more-chinese-turn-to-smartphone-program.html?_r=1.

Page 118   Levy, David, *Love + Sex with Robots: The Evolution of Human-Robot Relations* (New York: HarperCollins, 2007).

Page 119   Hempel, Jessi, "Computers That Know How You Feel Will Soon Be Everywhere," *Wired*, April 22, 2015: wired.com/2015/04/computers-can-now-tell-feel-face/.

Chapter 5:
How AI Put Our Jobs in Jeopardy

Page 123   Baker, Stephen, *Final Jeopardy: Man vs. Machine and the Quest to Know Everything* (Boston: Houghton Mifflin Harcourt, 2011).

Page 129   Keynes, John Maynard, "Economic Possibilities for Our Grandchildren," 1930: econ.yale.edu/smith/econ116a/keynes1 .pdf.

Page 130   Clark, Gregory, *A Farewell to Alms: A Brief Economic History of the World* (Princeton: Princeton University Press, 2007).

Page 130   Wong, Tessa, "Drone Waiters to Plug Singapore's Service Staff Gap," BBC News, February 8, 2015: bbc.co.uk/news/world-asia -31148450.

Page 131   Ford, Martin, *Rise of the Robots: Technology and the Threat of a Jobless Future* (New York: Basic Books, 2015): salon.com/2015/05/ 10/robots_are_coming_for_your_job_amazon_mcdonalds_and _the_next_wave_of_dangerous_capitalist_disruption/.

Page 131   Manyika, James et al. "Disruptive Technologies: Advances That Will Transform Life, Business, and the Global Economy," McKinsey Global Institute, May 2013: mckinsey.com/insights/ business_technology/disruptive_technologies.

Page 131   McCulloch, Warren, "Why the Mind Is in the Head?," *Dialectica*, vol. 4, no. 3, 1950.

Page 133   "Google's Self-Driving Car in Five-Car Crash," *Mountain View Voice*, August 8, 2011.

Page 133   Black, Thomas, "Google Sees Automated Flight Going 'All the Way' to airlines," *Bloomberg*, May 5, 2015: https://www .bloomberg.com/news/articles/2015–05-05/google-sees -automated-flight-going-all-the-way-to-airliners.

Page 133   Frey, Carl and Osborne, Michael, "The Future of Employment: How Susceptible Are Jobs to Computerisation?," Oxford Martin School, September 17, 2013: fhi.ox.ac.uk/wp-content/uploads/ The-Future-of-Employment-How-Susceptible-Are-Jobs-to -Computerization.pdf.

Page 134   Wile, Rob, "A Venture Capital Firm Just Named an Algorithm to Its Board Of Directors—Here's What It Actually Does,"

Business Insider, May 13, 2015: http://www.businessinsider
.com/vital-named-to-board-2014–5?IR=T.

Page 135   Kellenbenz, Herman, "Technology in the Age of the Scientific
Revolution, 1500–1700," in C.M. Cipolla (ed.) *The Fontana
Economic History of Europe*, vol 2. (London, Collins/Fontana,
1970).

Page 136   ruchalachimney.com/history.html.

Page 137   Wilde, Oscar, "The Soul of Man Under Socialism," 1891.

Page 137   Dormehl, Luke, "Foxconn CEO Unimpressed by his iPhone-
Building Robot Army," *Cult of Mac*, September 22, 2014:
cultofmac.com/297110/foxconn-ceo-unimpressed-iphone
-building-robot-army/.

Page 138   Mokdad, Ali et al., "Actual Causes of Death in the United States,
2000," American Medical Association, 2004: csdp.org/
research/1238.pdf.

Page 140   http://arstechnica.co.uk/gaming/2015/07/pewdiepie-responds-to
-haters-over-his-4–5-million-youtube-earnings/.

Page 141   http://uk.ign.com/articles/2013/10/09/gta-5-currently-holds
-seven-guinness-world-records.

Page 141   Gaudiosi, John, "New Reports Forecast Global Video Games
Industry Will Reach $82 Billion by 2017," *Forbes*, July 18, 2012:
http://www.forbes.com/sites/johngaudiosi/2012/07/18/
new-reports-forecasts-global-video-game-industry-will-reach
-82-billion-by-2017/.

Page 141   http://www.pwc.co.uk/en_uk/uk/assets/pdf/ukeo-regional
-march-2015.pdf.

Page 142   "Artificial Artificial Intelligence," *The Economist*, June 8, 2006:
economist.com/node/7001738?story_id=7001738.

Page 142   Chen, Edwin, "Improving Twitter Search with Real-Time
Human Computation," January 8, 2013: http://blog.echen
.me/2013/01/08/improving-twitter-search-with-real-time-human
-computation.

Page 143   forums.aws.amazon.com/thread.jspa?threadID=58891.

Page 143   Cushing, Ellen, "Amazon Mechanical Turk: The Digital
Sweatshop," *East Bay Express*, Jan/Feb 2013: utne.com/
science-and-technology/amazon-mechanical-turk
-zm0z13jfzlin.aspx.

Page 144   Conley, Neil and Braegelmann, Tom, "Decision of the German

Federal Supreme Court no. I ZR 112/06," *Journal of the Copyright Society*, Vol. 56, 2009: http://papers.ssrn.com/sol3/papers.cfm?abstract_id=1504982.

Page 144   Lanier, Jaron, *Who Owns the Future?* (New York: Simon & Schuster, 2013).

Page 146   Love, Julia, "Apple Ups Its AI Experts Hiring, But Faces Obstacles . . . ," *Reuters*, September 7, 2015: venturebeat .com/2015/09/07/apple-ups-its-a-i-hiring-but-faces-obstacles-to -making-phones-smarter/.

Page 147   Dredge, Stuart, "Apple Music Interview," *Guardian*, June 9, 2015: theguardian.com/technology/2015/jun/09/apple-music -interview-jimmy-iovine-eddy-cue.

Page 147   Dormehl, Luke, "New Apple Patents Hint at a Headphone, Music Revolution," *Cult of Mac*, May 29, 2014: cultofmac.com/281353/ apple-focus-reinventing-headphones/.

Page 149   Moss, Caroline, "Meet the Guy Who Makes $1,000 an Hour Tutoring Kids of Fortune 500 CEOs Over Skype," *Business Insider*, August 26, 2014: http://www.businessinsider.com /anthony-green-tutoring-2014–8?IR=T.

Page 149   Tabuchi, Hiroko, "Etsy's Success Gives Rise to Problems of Credibility and Scale," *New York Times*, March 15, 2015: nytimes.com/2015/03/16/business/media/etsys-success-raises -problems-of-credibility-and-scale.html?_r=0.

Chapter 6:
Can AI Be Creative?

Page 153   Mordvintsev, Alexander and Tyka, Mike, "Inceptionism: Going Deeper into Neural Networks," Google Research Blog, June 17, 2015: http://googleresearch.blogspot.co.uk/2015/06/ inceptionism-going-deeper-into-neural.html.

Page 156   Jefferson, Geoffrey, "The Mind of Mechanical Man," *British Medical Journal*, Vol. 1, June 25, 1949.

Page 157   Smith, Sylvia, "Iamus: Is this the 21st Century's Answer to Mozart?," BBC News, January 3, 2013: bbc.co.uk/news/ technology-20889644.

Page 158   Li, Boyang and Riedl, Mark, "Scheherazade: Crowd-Powered

Interactive Narrative Generation," Proceedings of the 29th AAAI Conference on Artificial Intelligence, 2015: cc.gatech .edu/~riedl/pubs/aaai15.pdf.

Page 163   There was only one ostensible mistake: *Abbey Road* was actually released in September 1969, around eight months before *Let It Be*, which arrived in May 1970. However, as any Beatles fan knows, the majority of the songs on *Let It Be* were recorded at the start of 1969, before *Abbey Road*—therefore making Shamir's algorithm correct.

Page 165   Behrens, Sam, "How Creepy Holographic Concerts Are Transforming the Future of the Music Business," *Music Mic*, May 22, 2014: mic.com/articles/89785/how-creepy-holographic -concerts-are-transforming-the-future-of-the-music-business# .AJp0Deyhu.

Page 167   Hobbes, Thomas, *Of Man: Being the First Part of Leviathan* (New York: Bartleby, 2001).

Page 168   Harry, Bill, *The Beatles Encyclopedia* (London: Virgin, 2000).

Page 170   Jackson, Joab, "Google: 129 Million Different Books Have Been Published," IDG News Service: pcworld.com/article/202803/ google_129_million_different_books_have_been_published .html.

Page 170   Johnson, Simon, "Britain's Most Avid Reader, 91, Has Borrowed 25,000 Library Books," *Telegraph*, June 29, 2009: telegraph.co .uk/news/newstopics/howaboutthat/5932159/Britains-most -avid-reader-91-has-borrowed-25000-library-books.html.

Page 175   Stevenson, Mark, *An Optimist's Tour of the Future* (London: Profile, 2011).

Page 175   Geere, Duncan, "Genetic Algorithms Find Unbeatable StarCraft Tactics," *Wired*, November 2, 2010: wired.com/2010/11/ genetic-algorithms-starcraft/.

Page 178   Watson, Chef, *Cognitive Cooking with Chef Watson* (Naperville, Illinois: Sourcebooks, 2015).

Chapter 7:
In the Future There Will Be Mindclones

Page 183   Thankfully for those of us over thirty, Jobs went on to disprove
           his own hypothesis by helping create the iPod, iPhone and iPad
           well into his fifties.

Page 188   Kopalle, Praveen, "Why Amazon's Anticipatory Shipping Is
           Pure Genius," *Forbes*, January 28, 2014: forbes.com/sites/
           onmarketing/2014/01/28/why-amazons-anticipatory-shipping
           -is-pure-genius/#664afc342fac.

Page 189   Kelion, Leo, "CES 2014: Sony Shows Off Life Logging App and
           Kit," BBC News, January 7, 2014: bbc.co.uk/news/
           technology-25633647.

Page 191   Landauer, Thomas, "How Much Do People Remember? Some
           Estimates of the Quantity of Learned Information in Long-
           Term Memory," *Cognitive Science*, Vol. 10, 1986.

Page 192   cmu.edu/homepage/computing/2010/summer/synthetic-abe
           .shtml.

Page 195   Minsky, Marvin, "Will Robots Inherit the Earth?," *Scientific
           American*, October 1994: http://web.media.mit.edu/~minsky/
           papers/sciam.inherit.html.

Page 197   Triantaphyllou, Evangelos, *Data Mining and Knowledge
           Discovery via Logic-Based Methods: Theory, Algorithms, and
           Applications* (New York: Springer, 2010).

Page 197   Hsu, Jeremy, "Biggest Neural Network Ever Pushes AI Deep
           Learning," *Spectrum*, July 8, 2015: spectrum.ieee.org/tech-talk/
           computing/software/biggest-neural-network-ever-pushes-ai
           -deep-learning.

Page 198   Randerson, James, "How Many Neurons Make a Human Brain?
           Billions Fewer than We Thought," *Guardian*, February 28, 2012:
           theguardian.com/science/blog/2012/feb/28/how-many-neurons
           -human-brain.

Page 200   Stoller-Conrad, Jessica, "Controlling a Robotic Arm with a
           Patient's Intentions," Caltech, May 21, 2015: caltech.edu/news/
           controlling-robotic-arm-patients-intentions-46786.

Page 201   Kever, Jeannie, "Researchers Build Brain-Machine Interface to
           Control Prosthetic Hand," University of Houston, March 31,

2015: uh.edu/news-events/stories/2015/March/0331BionicHand
.php.

Page 202   Kurzweil, Ray, "The Law of Accelerating Returns," March 7,
            2001: kurzweilai.net/the-law-of-accelerating-returns.

Page 202   Linden, David, "The Singularity Is Far: A Neuroscientist's
            View," *BoingBoing*, July 14, 2011: http://boingboing.net/2011/
            07/14/far.html.

Page 206   http://2045.com/press/.

Page 206   Hayworth, Ken, "Killed by Bad Philosophy," Brain Preservation
            Foundation, January 2010: brainpreservation.org/content-2/
            killed-bad-philosophy/.

Chapter 8:
The Future (Risks) of Thinking Machines

Page 211   Cook, James, "Elon Musk: Robots Could Start Killing Us All
            Within 5 Years," *Business Insider*, November 17, 2014: http://uk
            .businessinsider.com/elon-musk-killer-robots-will-be-here
            -within-five-years-2014-11.

Page 211   Hern, Alex, "Elon Musk Says He Invested in DeepMind Over
            'Terminator' Fears," *Guardian*, June 18, 2014: theguardian.com/
            technology/2014/jun/18/elon-musk-deepmind-ai-tesla-motors.

Page 212   Hawking, Stephen et al., "Stephen Hawking: 'Transcendence
            Looks at the Implications of Artificial Intelligence . . . ,' "
            Independent, May 1, 2014: independent.co.uk/news/science/
            stephen-hawking-transcendence-looks-at-the-implications-of
            -artificial-intelligence-but-are-we-taking-9313474.html.

Page 215   Hill, Doug, "The Eccentric Genius Whose Time May Have Finally
            Come (Again)," *Atlantic*, June 11, 2014: theatlantic.com/technology/
            archive/2014/06/norbert-wiener-the-eccentric-genius-whose-time
            -may-have-finally-come-again/372607/.

Page 215   Good, I. J., "Speculations Concerning the First Ultraintelligent
            Machine," *Advances in Computers*, 1965.

Page 216   Vinge, Vernor, "The Coming Technological Singularity: How to
            Survive in the Post-Human Era," *Vision-21: Interdisciplinary
            Science and Engineering in the Era of Cyberspace*, 1993.

Page 217   Ulam, Stanislaw, "Tribute to John von Neumann," *Bulletin of the American Mathematical Society*: 5, May 1958.

Page 218   Hemingway, Ernest, *The Sun Also Rises* (New York: Scribner, 1954).

Page 218   Appleyard, Bryan, *The Brain Is Wider than the Sky: Why Simple Solutions Don't Work in a Complex World* (London: Weidenfeld & Nicholson, 2011).

Page 222   www.youtube.com/watch?v=KbOqsp3oUQI.

Page 223   Bostrom, Nick, *Superintelligence: Paths, Dangers, Strategies* (Oxford, UK: Oxford University Press, 2014).

Page 224   Steiner, Christopher, *Automate This: How Algorithms Came to Rule Our World* (New York: Portfolio/Penguin, 2012).

Page 225   Bostrom, Nick and Eliezer Yudkowsky, "The Ethics of Artificial Intelligence," *Cambridge Handbook of Artificial Intelligence*, 2011: nickbostrom.com/ethics/artificial-intelligence.pdf.

Page 227   opencog.org.

Page 228   Cole, Dermot, "iPhone Map App Directs Fairbanks Drivers onto Airport Taxiway," *Alaska Dispatch News*, September 24, 2013: adn.com/article/20130924/iphone-map-app-directs -fairbanks-drivers-airport-taxiway.

Page 229   Vladeck, David, "Machines Without Principals: Liability Rules and Artificial Intelligence," *Washington Law Review*, vol. 89, March 2014.

Page 232   Mizroch, Amir, "Google on Artificial Intelligence Panic: Get a Grip," *WSJ*, June 8, 2015: blogs.wsj.com/digits/2015/06/08/google-on -artificial-intelligence-panic-get-a-grip/.

Page 235   Thompson, Bill, "The Cruelty of the First Law of Robotics," August 13, 2015: astickadogandaboxwithsomethinginit .com/2015/08/the-cruelty-of-the-first-law-of-robotics/.

Page 235   Adams, Douglas, *The Restaurant at the End of the Universe* (New York: Harmony Books, 1980).

Page 236   Descartes, René, *Descartes: Selected Philosophical Writings* (New York: Cambridge University Press, 1988).

Conclusion:
Rise of the Robots

Page 239   Rifkin, Glenn, "Marvin Minsky, Pioneer in Artificial
            Intelligence, Dies at 88," *New York Times*, January 25, 2015:
            nytimes.com/2016/01/26/business/marvin-minsky-pioneer-in
            -artificial-intelligence-dies-at-88.html?_r=1.

Page 239   Rogers, Adam, "We Asked a Robot to Write an Obit for AI
            Pioneer Marvin Minsky," *Wired*, January 26, 2016: wired
            .com/2016/01/we-asked-a-robot-to-write-an-obit-for-ai-pioneer
            -marvin-minsky/.

Page 240   Minsky, Marvin, *Society of Mind* (New York: Simon and
            Schuster, 1986).

Page 240   HAL 90210, "No Go: Facebook Fails to Spoil Google's Big AI
            Day," *Guardian*, January 28, 2016: theguardian.com/
            technology/2016/jan/28/go-playing-facebook-spoil-googles-ai
            -deepmind.

Page 241   Moyer, Christopher, "How Google's AlphaGo Beat a Go World
            Champion," *Atlantic*, March 28, 2016: http://www.theatlantic
            .com/technology/archive/2016/03/the-invisible-opponent
            /475611.

Page 241   "US Military Shelves Google Robot Plan Over 'Noise
            Concerns,'" BBC News, December 30, 2015: bbc.co.uk/news/
            technology-35201183.

Page 242   Collins, Ben, "Meet the Robot Writing 'Friends' Sequels," *Daily
            Beast*, January 20, 2016: thedailybeast.com/articles/2016/01/20/
            meet-the-robot-writing-friends-sequels.html.

INDEX

EAGLE VALLEY LIBRARY DISTRICT
P.O. BOX 240 600 BROADWAY
EAGLE, CO 81631 / 328-8800

ABOUT THE AUTHOR

Luke Dormehl is a British technology writer, public speaker and author of *The Apple Revolution* and *The Formula: How Algorithms Solve All Our Problems . . . And Create More.* He has written for *Wired*, the *Guardian*, *Fast Company*, the *Sunday Times* and others, as well as directing a range of television documentaries.

EAGLE VALLEY LIBRARY DISTRICT
P.O. BOX 240 600 BROADWAY
EAGLE, CO 81631 / 328-8800

EAGLE VALLEY LIBRARY DISTRICT
P.O. BOX 240 600 BROADWAY
EAGLE, CO 81631 / 328-8800

ABOUT THE AUTHOR

Luke Dormehl is a British technology writer, public speaker and author of *The Apple Revolution* and *The Formula: How Algorithms Solve All Our Problems . . . And Create More.* He has written for *Wired*, the *Guardian*, *Fast Company*, the *Sunday Times* and others, as well as directing a range of television documentaries.

EAGLE VALLEY LIBRARY DISTRICT
P.O. BOX 240 600 BROADWAY
EAGLE, CO 81631 / 328-8800